Lecture Notes in Mathematics

Edited by A. Dold, Heidelberg and B. Eckmann, Zürich

377

A. M. Fink

Iowa State University of Science and Technology, Ames, IA/USA

Almost Periodic Differential Equations

Springer-Verlag
Berlin · Heidelberg · New York 1974

AMS Subject Classifications (1970): 34-02, 34-C-25, 42-02, 42-A-84, 34-C-30, 34-D-20

ISBN 3-540-06729-9 Springer-Verlag Berlin · Heidelberg · New York
ISBN 0-387-06729-9 Springer-Verlag New York · Heidelberg · Berlin

Offsetdruck: Julius Beltz, Hemsbach/Bergstr.

Preface

These lecture notes are written with the hope that the recent advances in the subject of almost periodic differential equations can become more accessible if the basic facts are collected in one place. It is well known that in Celestial Mechanics, almost periodic solutions and stable solutions are intimately related. In the same way, stable electronic circuits exhibit almost periodic behavior. A vast amount of research has been directed toward studying these phenomena. A great portion of the roughly five hundred items in our bibliography are dated after 1955.

These lecture notes are about almost periodic solution to ordinary differential equations. I spend the first four chapters on the theory of almost periodic functions. Included in those chapters is the skeleton of almost periodic theory. I have taken the tack of presenting only that material which is germane to the later chapters on differential equations. I include essentially no fact about almost periodic functions which is not used to prove something else. This is no virtue. It illustrates the depth of the thoery that is developed here.

These notes are self-contained except for the usual preliminary facts about existence and uniqueness of solutions of differential equations. It should therefore be accessible to a wide audience. I have taken the periodic case as motivation for much of the material, so an acquaintance with the fourier series theory of periodic functions is helpful.

Much of the material in the first four chapters appears in other books. Some of the material, however, is only available in research papers. This is primarily the case because the theory of almost periodic differential equations has been a rich source for new developments in almost periodic functions that are not well-known.

I have tried to give references to specific results. The reader will find an extensive bibliography of the subject. In addition, the notes at the end of each chapter are helpful in identifying sources of results as well as sources for a study of the literature.

The table of contents and the index are reasonably extensive and should be useful to the casual reader.

It is obvious that these lecture notes can only be an introduction to the subject. The number of items in the bibliography attests to that. Much interesting material is not mentioned. Topological dynamics is not included. It would require a book in itself, see Jacobs [546]. A specialist in the subject may easily find some favorite idea that is missing. I only hope that many readers can find something to tickle their imagination.

These notes have been used in various forms, for seminars at the University of Nebraska, the University of Colorado, and Iowa State University. The participants in those seminars have materially aided the exposition given here. In particular, my colleague, George Seifert, has read these notes and his comments, when followed, have improved the exposition. I thank him for his efforts.

Table of Contents

Chapter 1

Almost Periodic Functions

1. Introduction. There are several known equivalent definitions of almost periodic functions. Which one to select as the basic definition is therefore a choice which depends on the inclination of the writer. Our choice here is to select the one that is most useful in its applications to differential equations. In doing this, we reverse somewhat the history of the subject.

We shall shortly be concerned with vector valued functions but for our first considerations, we restrict our attention to complex valued functions of a real variable. Furthermore, we assume that all functions considered in this book are continuous and make no further note of the matter.

2. The definition. We propose to use the definition that uses sequential convergence.

Definition 1.1. The function f is almost periodic if from every sequence $\{\alpha_n'\}$ one can extract a subsequence $\{\alpha_n\}$ such that $\lim_{n\to\infty} f(t+\alpha_n)$ exists uniformly on the real line.

We will want to make some basic observations before we proceed. Almost periodic functions are intended to be generalizations of periodic functions in some sense. It is clear that periodic functions satisfy Definition 1.1. Indeed, if f has period T, then from the given sequence $\{\alpha_n'\}$ we may select a subsequence $\{\alpha_n\}$ such that $\{\alpha_n (\mathrm{mod}\ T)\}$ converges to α_0.

Then $\lim_{n\to\infty} f(t+\alpha_n) = f(t+\alpha_0)$. Thus the almost periodicity of f, for periodic f, is a statement about the compactness of $A = \{f_\tau : f_\tau(t) = f(t+\tau)\}$ in the space of continuous functions with the uniform norm. For general almost periodic functions we replace compactness by pre-compactness or by A having compact closure.

To make these remarks precise we want to introduce some useful notation. As a preliminary, note that almost periodic functions are bounded. In fact, if $|f(\alpha_n')| \to +\infty$ then for no subsequence can $\{f(t+\alpha_n)\}$ converge at 0.

<u>Notations 1.2</u>. Let $\|f\| = \sup_t |f(t)|$ and $AP(C) \equiv \{f : f$ is an almost periodic complex valued function$\}$ and we endow this space with the norm $\|\ \|$, C designates the complex numbers, R the real numbers, and $AP(R)$ is similary defined. $AP(E)$ is intended to mean either one. Secondly, we want to avoid the necessity for double subscripts when taking subsequences.

<u>Notation 1.3</u>. The sequence $\{\alpha_n\}$ is denoted by α. If $\beta = \{\beta_n\}$ then $\beta \subset \alpha$ means that β is a subsequence of α; $\alpha + \beta = \{\alpha_n + \beta_n\}$; $-\alpha = \{-\alpha_n\}$; and α and β are common subsequences of α' and β' respectively means that $\alpha_n = \alpha'_{n(k)}$ and $\beta_n = \beta'_{n(k)}$ for some given function $n(k)$.

In order to ease the notation for taking limits we introduce the translation operator T.

Notation 1.4. $T_\alpha f = g$ means that $g(t) = \lim_n f(t + \alpha_n)$ and is written only when the limit exists. The mode of convergence, e.g. pointwise, uniform, etc., will be specified at each use of the symbol.

Using these notations, the definition of almost periodic function becomes: f is almost periodic if for every α', there is $\alpha \subset \alpha'$ such that $T_\alpha f$ exists uniformly.

3. The hull. An important notion in almost periodic differential theory is the hull.

Definition 1.5. $H(f) \equiv \{g \mid \text{there exists } \alpha \text{ with } T_\alpha f = g$ uniformly$\}$, is called the hull of f.

We have noted above that if f is periodic then $H(f) \equiv \{f_\tau : f_\tau(t) = f(t + \tau), \tau \in R\}$ and is compact as a subset of $AP(C)$. For general almost periodic functions, there are functions in $H(f)$ which are not translates of f. As an example, we will show below that $f(t) = \cos t + \cos\sqrt{2}\, t$ is almost periodic. Note that $f(t) > -2$ for all t and there is a sequence α'_n such that $f(\alpha'_n) \to -2$. If $\alpha \subset \alpha'$ such that $g = T_\alpha f$ uniformly, then $g(0) = -2$. Thus g is not a translate of f. It is this phenomena which makes almost periodic theory interesting and at the same time difficult. If the reader persists until Chapter 6, he will fully appreciate this statement.

Theorem 1.6. $H(f)$ is compact in uniform norm if and only if f is almost periodic.

Proof: If $H(f)$ is compact, then certainly given α' there is an $\alpha \subset \alpha'$ such that $T_\alpha f$ exists uniformly. Conversely, if f is almost periodic and $\{g_n\}$ is a sequence from $H(f)$, then we pick α'_n such that $\|f(t+\alpha'_n) - g_n(t)\| < 1/n$. Find $\alpha \subset \alpha'$ so that $T_\alpha f$ exists uniformly. Let $\beta \subset \gamma = \{n\}$ so that β and α are common subsequences, then $f(t+\alpha_n) - g_{\beta_n}(t) \to 0$ so that $g_{\beta_n} \to T_\alpha f$.

Since we are considering $H(f)$ as a metric space one sees that the compactness is equivalent to the fact that the set of translates $\{f(t+\tau)\}_{\tau\in R}$ is totally bounded. Indeed, this is the definition adopted by von Neumann and essentially used by Maak in his treatise. These remarks prove the next theorem.

Theorem 1.7. The function f is almost periodic if and only if for every $\epsilon > 0$, there are numbers $a_1,\ldots,a_n$ and a function $n(t)$ from R to $\{1,\ldots,n\}$ such that
$|f(t+a_{n(\tau)}) - f(t+\tau)| < \epsilon$ for all t and τ.

We will not have much occasion to use this result, but the next one is crucial.

Theorem 1.8. If f is almost periodic, then for any $g \in H(f)$, $H(g) = H(f)$.

Proof: If α is given with $T_\alpha g = h$ uniformly and $g \in H(f)$, then select β so that $|f(t+\beta_n) - g(t+\alpha_n)| < 1/n$. It follows that $T_\beta f = h$ so that $h \in H(f)$. Thus $H(g) \subseteq H(f)$. On the other hand, if $T_\alpha f = g$ then $|f(t+\alpha_n) - g(t)| \to 0$ so making the change of variable $t + \alpha_n \to s$ one has

$|f(s) - g(s-\alpha_n)| \to 0$, i.e. $T_{-\alpha}g = f$. Thus $f \in H(g)$ and so by what was shown above, $H(f) \subseteq H(g)$.

4. <u>The space AP(C)</u>. We have observed that one class of functions that is in AP(C) is the periodic functions. If one considers the set of all periodic functions with the period not fixed, this set is not very nice. In particular, it is not closed under sums or uniform limits. The space AP(C) remedies both of these defects.

<u>Theorem 1.9</u>. AP(C) is an algebra over C. It is closed under conjugation, the lattice operations on real functions, and uniform limits. Furthermore, if $f \in AP(C)$ and F is uniformly continuous on the range of f to the complex numbers, then $F \circ f \in AP(C)$. If $f \in AP(C)$ and $\inf_t |f(t)| > 0$, then $\frac{1}{f(t)} \in AP(C)$.

<u>Proof</u>: It is clear that AP(C) is closed under taking scalar multiples. If f and $g \in AP(C)$ and α'' is a given sequence, find $\alpha' \subset \alpha''$ so that $T_{\alpha'}f$ exists uniformly. Then find $\alpha \subset \alpha'$ so that $T_\alpha g$ exists uniformly. Then $T_\alpha(f+g) = T_\alpha f + T_\alpha g = T_{\alpha'}f + T_\alpha g$ exists uniformly. Thus $f + g \in AP(C)$. Clearly now, any finite sum of elements of AP(C) is also in AP(C). As above if $\alpha \subset \alpha''$ so that $T_\alpha f$ and $T_\alpha g$ exist uniformly, then $T_\alpha(fg) = (T_\alpha f)(T_\alpha g)$ also exists uniformly so that AP(C) is closed under products. Thus AP(C) is an algebra.

If $f \in AP(C)$ and F is uniformly continuous on the range

of f then $T_\alpha(F \circ f) = F(T_\alpha f)$ so that $F \circ f \in AP(C)$. In particular, taking $F(z) = |z|$ we get $AP(C)$ closed under absolute values and for the choice $F(z) = \overline{z}$ closed under conjugation. If f and g are real valued functions in $AP(C)$, then $\min(f,g) = \dfrac{(f+g) - |f-g|}{2} \in AP(C)$ by the above observations, as is $\max(f,g) = \dfrac{(f+g) + |f-g|}{2}$. Furthermore, if $\inf_t |f(t)| > 0$ then $\frac{1}{z}$ is uniformly continuous on the range of f, so $\frac{1}{f(t)} \in AP(C)$ if f is. To get the property of being closed under uniform limits one could use a diagonal process, however it is much easier to use the characterization of Theorem 1.7. For let $f_n \in AP(C)$ for all n and $f_n \to f$ uniformly. If $\epsilon > 0$ is given, find m so that $\|f_m - f\| < \epsilon/3$, and for f_m find the numbers $a_1,\ldots,a_n$ as in Theorem 1.7 as well as the function $n(\tau)$ for ϵ replaced by $\epsilon/3$. Then

$$|f(t+\tau) - f(t+n(\tau))| \le |f(t+\tau) - f_m(t+\tau)|$$
$$+ |f_m(t+\tau) - f_m(t+n(\tau))| + |f_m(t+n(\tau)) - f(t+n(\tau))| < \epsilon.$$

Thus f satisfies Theorem 1.7 and is in $AP(C)$.

Note now that we have shown that sums (finite) of periodic functions are almost periodic functions. In particular $\cos t + \cos\sqrt{2}t \in AP(C)$, as was previously noticed. This function is also a simple example that shows that the sum of two periodic functions need not be periodic. In fact, the equation $\cos t + \cos\sqrt{2}t = 2$ has precisely one solution so that this function is not periodic.

Also note that a uniformly convergent series of periodic functions is in AP(C) since the partial sums are in AP(C) by its algebra properties. Thus AP(C) is a complete normed space of functions that contains all periodic functions. In fact, it is the smallest complete normed space with this property. The proof of this fact will be postponed. To be perfectly honest, since the topic of this book is differential equations, one ought to immediately ask whether AP(C) is closed under differentiation and integration. Roughly speaking it is under the minimum hypothesis, but we again postpone these considerations to more appropriate circumstances.

5. The Bohr Definition. As we have remarked, we have inverted some history in introducing the Bochner definition first. We have done this precisely for the reason that Theorem 1.9 is most easily proved by this definition. However, the fourier series theory is most easily established using Bohr's definition.

Definition 1.10. A subset S of R is called relatively dense if there exists a positive number L such that $[a,a+L] \cap S \neq \phi$ for all $a \in R$. The number L is called the inclusion length.

Roughly speaking, the complement of S should not contain arbitrarily long intervals. Thus any doubly infinite arithmetic sequence is relatively dense.

Definition 1.11. For any bounded complex function f and $\epsilon > 0$, we define $T(f,\epsilon) \equiv \{\tau \mid |f(t+\tau) - f(t)| < \epsilon \text{ for all } t\}$.

$T(f,\epsilon)$ is called the ϵ-translation set of f.

As two simple examples, we note that uniform continuity of f is equivalent to the inclusion $(-\delta,\delta) \subset T(f,\epsilon)$ for some function $\delta(\epsilon) > 0$. Also if f is periodic, then one easily sees that $\bigcup_{n=-\infty}^{\infty} (-\delta(\epsilon) + nT, \delta(\epsilon) + nT) \subset T(f,\epsilon)$ where T is the period of f. In fact, it is easy to verify that for $f(t) = \sin t$ and ϵ sufficiently small, there is a continuous function $\delta(\epsilon)$ for which equality holds. In any case, for periodic functions it is clear that $T(f,\epsilon)$ is relatively dense since the functions $f(t+nT)$ fit exactly on top of $f(t)$. The geometric significance of $T(f,\epsilon)$ is a generalization of this idea, where the translated graph comes within ϵ of the graph of f if $\tau \in T(f,\epsilon)$.

Definition 1.12. We say f is (Bohr) almost periodic if for every $\epsilon > 0$, $T(f,\epsilon)$ is relatively dense.

The major result here is that the definitions 1.1 and 1.12 lead to the same class of functions and so the adjective "Bohr" may be dropped in Definition 1.12.

Theorem 1.13. A Bohr almost periodic function is uniformly continuous.

Proof: If $\epsilon > 0$, let L be the inclusion interval for $T(f,\epsilon)$ and $\delta > 0$ so that $|t-s| < \delta < 1$, and $t,s \in [-1,L+1]$ imply that $|f(t)-f(s)| < \epsilon$. If τ and σ are real numbers so that $|\tau-\sigma| < \delta$ and x is a real number we pick $t \in T(f,\epsilon) \cap [-\tau-x, -\tau-x+L]$. Then $x+\tau+t \in [0,L]$,

$x + \tau + \sigma \in [-1, L+1]$ and thus

$|f(x+\tau) - f(x+\sigma)| \le |f(x+\tau) - f(x+\tau+t)|$

$+ |f(x+\tau+t) - f(x+\tau+\sigma)| + |f(x+\tau+\sigma) - f(x+\sigma)| < 3\epsilon$ since $t \in T(f,\epsilon)$ and $|(x+\tau+t) - (x+\tau+\sigma)| < \delta$. This holds for all x so f is uniformly continuous.

The above proof exemplifies the idea of almost periodic function theory that everything that happens, almost happens on some finite interval. This method is also illustrated in the next theorem.

Theorem 1.14. For any complex function f, $f \in AP(C)$ if and only if f is Bohr almost periodic.

Proof: If f is Bohr almost periodic then according to Theorem 1.13, f is uniformly continuous. If $\epsilon > 0$ is given, then choose $\delta > 0$ so that $|t-s| < \delta$ implies $|f(x+t) - f(x+s)| < \epsilon/2$ for all x. Let $a_i = i\delta$ $i = 1,\dots,n$ where $n\delta > L$ and L is the inclusion interval of $T(f,\epsilon/2)$. If $\tau \in R$ then there is a number $\sigma \in [-\tau, -\tau+L] \cap T(f,\epsilon/2)$. Thus $|f(x+\sigma) - f(x)| < \epsilon/2$ for all x and $\sigma + \tau \in [0,L]$. There is a number $m(\tau)$ such that $|a_m - (\tau+\sigma)| < \delta$. Thus $|f(x+a_m) - f(x+\tau+\sigma)| < \epsilon/2$ for all x. Consequently, $|f(x+a_m) - f(x+\tau)| \le |f(x+a_m) - f(x+\tau+\sigma)|$ $+ |f((x+\tau)+\sigma) - f(x+\tau)| < \epsilon$, and f satisfies 1.7.

For the converse, if f is in $AP(C)$ and $\epsilon > 0$. As in 1.7 pick $a_1,\dots,a_n$ and $i(\tau)$ so that for all t and τ

$|f(t+a_{i(\tau)}) - f(t+\tau)| < \epsilon$. This is equivalent to $|f(t) - f(t+\tau - a_{i(\tau)})| < \epsilon$ for all t. Let $L = \max|a_i|$. Then $\tau - L \leq \tau - a_{i(\tau)} \leq \tau + L$, $\tau - a_{i(\tau)} \in T(f,\epsilon)$, and $2L$ is an inclusion length.

Corollary 1.15. Almost periodic functions are uniformly continuous.

6. Derivatives of AP functions. If $f \in AP(C)$ and f' exists everywhere is $f' \in AP(C)$? The answer is very simple. Since all functions in $AP(C)$ are bounded and uniformly continuous this is a necessary condition. In fact uniform continuity is sufficient.

Theorem 1.16. If $f \in AP(C)$ and f' exists then $f' \in AP(C)$ if and only if f' is uniformly continuous.

Proof. We need to show that if f' is uniformly continuous on R then $f' \in AP(C)$ if $f \subset AP(C)$. Now $f_n(t) = n[f(t+1/n) - f(t)] \to f'(t)$ pointwise. However, the convergence is uniform. In fact if $|f'(t) - f'(s)| < \epsilon$ when $|t-s| < \delta$, then pick $n \geq \delta^{-1}$. Then $f'(t) - f_n(t) = f'(t) - f'(t+\theta(t,n))$ by the mean value theorem, and $t \leq \theta(t,n) \leq t + 1/n \leq t + \delta$. Thus $|f'(t) - f_n(t)| < \epsilon$. Clearly $f_n \in AP(C)$ for each n so that $f' \in AP(C)$ by Theorem 1.9.

7. A pointwise definition. As we have seen, the Bochner definition 1.1 is most convenient for showing the algebraic and topological properties of AP(C). Looking ahead, when one wants to show that a solution of a differential equation is almost periodic, then this definition is hard to apply. The underlying reason for this is the non-compactness of the space on which the functions are defined. There are no general theorems of analysis which yield uniform convergence on R. Contrast this with the situation on a finite interval where the Arzela-Ascoli Theorem and Dini's Theorem give relatively nice sufficient conditions. Both these theorems obviously fail for general functions on R, but more, they also fail in AP(C). We are thus led to a state of affairs in which $T_\alpha f$ exists uniformly by virtue of apriori knowledge of its almost periodicity or the use of the definition of uniform convergence. Or we can resort to mathematical trickery! A reading of later chapters will show that this device meets with some success. In any case, much of this success is due to the observation that the topology of uniform convergence on compact subsets of R is very much nicer in some respect. That is, a sufficient condition for convergence of some subsequence of a given sequence in this topology is that the sequence consist of functions which are uniformly bounded and equicontinuous on R. We will illuminate this later. Suffice it to say, that what is needed is a version of almost periodicity that involves this kind of convergence.

Theorem 1.17. A function $f \in AP(C)$ if and only if from every pair of sequences α', β' there are common subsequences $\alpha \subset \alpha'$ and $\beta \subset \beta'$ such that

$$T_{\alpha+\beta} f = T_\alpha T_\beta f \qquad \text{pointwise.} \tag{1.1}$$

Proof: If $f \in AP(C)$ then given α', β' we may find a subsequence $\beta'' \subset \beta'$ so that $T_{\beta''} f = g$ uniformly. Now g is in $AP(C)$ so that if $\alpha'' \subset \alpha'$ common with β'', we can find $\alpha''' \subset \alpha''$ so that $T_{\alpha'''} g = h$ uniformly. Letting $\beta''' \subset \beta''$ common with α''', we can find common subsequences $\alpha \subset \alpha'''$, $\beta \subset \beta'''$ so that $T_{\alpha+\beta} f = k$ uniformly. Now note that $T_\beta f = g$ and $T_\alpha g = h$ uniformly since $\beta \subset \beta''$ and $\alpha \subset \alpha'''$. If $\epsilon > 0$ is given then for all t, $|k(t) - f(t+\alpha_n+\beta_n)| < \epsilon$, $|g(t) - f(t+\beta_n)| < \epsilon$, and $|h(t) - g(t+\alpha_n)| < \epsilon$ if n is sufficiently large. Thus $|h(t) - k(t)| \leq |h(t) - g(t+\alpha_n)| + |g(t+\alpha_n) - f(t+\alpha_n+\beta_n)| + |f(t+\alpha_n+\beta_n) - k(t)| < 3\epsilon$. Since ϵ is arbitrary, $h \equiv k$. Thus the condition is necessary with uniform convergence !

Conversely, suppose γ' is a sequence. Taking $\alpha' = 0$ and $\beta' = \gamma'$ we see that condition (1.1) yields $\gamma \subset \gamma'$ so that $T_\gamma f$ exists pointwise. Suppose, by way of contradiction, that the convergence is not uniform on R. There exist sequences $\alpha' \subset \gamma$ $\beta' \subset \gamma$ and t' and $\epsilon > 0$ so that

$$|f(\alpha_n' + t_n') - f(\beta_n' + t_n')| \geq \epsilon. \tag{1.2}$$

Apply the condition (1.1) to the sequences α' and t', to find common subsequences $\alpha'' \subset \alpha'$ and $t'' \subset t'$ (and $\beta'' \subset \beta'$

also common) so that pointwise

$$T_{\alpha''+t''}f = T_{t''}T_{\alpha''}f. \qquad (1.3)$$

Now apply the condition (1.1) to the sequences β'' and t'' to find common subsequences $\beta \subset \beta'$ and $t \subset t''$ (and again $\alpha \subset \alpha''$ common with them) so that pointwise

$$T_{t+\beta}f = T_t T_\beta f \qquad (1.4)$$

Now note that we also have via (1.3) that $T_{t+\alpha}f = T_t T_\alpha f$. But $T_\alpha f = T_\gamma f = T_\beta f$ since $\alpha \subset \gamma$ and $\beta \subset \gamma$. All this is pointwise. In any case we thus have $T_t T_\alpha f = T_t T_\beta f$ so that combining (1.3) and (1.4) we have $T_{t+\beta}f = T_{t+\alpha}f$. At zero this contradicts (1.2) since $t+\beta \subset t'+\beta'$ and $t+\alpha \subset t'+\alpha'$ and these are common subsequences. Thus the convergence is uniform.

Note the most remarkable thing. The condition (1.1) is in fact necessary and sufficient in any of the three modes of convergence, pointwise, uniform on compact subsets, and uniform. Of course the strongest version is that the condition (1.1) is necessary in the uniform sense and sufficient in the pointwise sense.

We can also note that Theorem 1.8 is an easy Corollary of Theorem 1. 17, for if $g = T_\alpha f$ then $T_\beta g = T_{\alpha+\beta}f$.

8. Notes. The definition of almost periodic functions 1.1 appears for the first time in Bochner, Beitrage zur Theorie der fastperiodische Funktionen. I : Funktionen einer Variablen, Math. Ann. 96(1927), 119-147. He showed its equivalence to the earlier Bohr definition. Bochner's definition makes sense in arbitrary groups so that its usefulness is certainly not limited to differential equations.

The notation used in 1.3 and 1.4 was also developed by Bochner in, A new approach to almost periodicity, Proc. Nat. Acad. Sci. USA 48(1962), 2039-2043. It is a very useful device.

The equivalence of pre-compactness and total boundedness can be found in any topology book.

The Bohr definition of course first appears in his original paper which is most easily found in the Collected Math. works, 1952. He shows that requiring only that $T(f,\epsilon)$ has arbitrarily large numbers leads to a different class of functions.

A different proof of Theorem 1.14 may be found in Besicovitch, Almost Periodic Functions, Dover, 1954.

Theorem 1.17 is due to Bochner and is found in the above cited paper of 1962. This paper also introduces the notion of almost automorphic function. If Theorem 1.17 is only applied to the case $\beta = -\alpha$ then one gets a different class of functions. Specifically, if from every α' we can extract $\alpha \subset \alpha'$ so that $T_{-\alpha}T_{\alpha}f = f$ pointwise, then f is almost automorphic. These functions have been studied by some students of Bochner and others. We mention the papers of Veech specifically. Many of the ideas discussed in later chapters yield the existence of almost automorphic solutions. It is an open question whether an almost periodic differential equation can have an almost automorphic solution that is not almost periodic. See Question H, page 168.

Chapter 2

Uniformly Almost periodic families

1. Introduction. We are aiming toward looking at the differential equation $x' = f(x,t)$ where f is an almost periodic function as a function of t and where x is considered a parameter. In a search for an almost periodic solution $\varphi(t)$ we will need to consider the composition $f(\varphi(t),t)$. Is this an almost periodic function? Surprisingly, the answer is in general, no! Consider $f(x,t) = \sin xt$, $x \in R$. For each x in R, f is periodic in t, so certainly it is almost periodic. Now consider $f(\sin t,t) = \sin(t \sin t)$. This is not almost periodic, in fact it is not uniformly continuous. If it were uniformly continuous then $\lim_{n\to\infty} \sin[(m\pi + \frac{1}{n})\sin(m\pi + \frac{1}{n})] = 0$ uniformly in m. But for $m = \frac{n}{2}$ and n even, one notes that $(\frac{n\pi}{2} + \frac{1}{n})\sin(n\pi + \frac{1}{n}) = (\frac{n\pi}{2} + \frac{1}{n})\sin\frac{1}{n}$ can be estimated by

$$(\frac{n\pi}{2} + \frac{1}{n})\sin\frac{1}{n} \leq (\frac{n\pi}{2} + \frac{1}{n})\frac{1}{n} = \frac{\pi}{2} + \frac{1}{n^2} \quad \text{and}$$

$$(\frac{n\pi}{2} + \frac{1}{n})\sin\frac{1}{n} \geq (\frac{n\pi}{2} + \frac{1}{n})\frac{2}{\pi n} = 1 + \frac{2}{\pi n^2}$$

so that

$$\sin 1 \leq \sin(1 + \frac{2}{\pi n^2}) \leq \sin[(m\pi + \frac{1}{n})\sin(m\pi + \frac{1}{n})]$$

for n even $m > \frac{n}{2}$ and n large.

With this example in mind, we see what sort of hypothesis is required. Let us start a proof. In terms of ϵ-translation numbers one would look at $f(\varphi(t+\tau), t+\tau) - f(\varphi(t), t) = f(\varphi(t+\tau), t+\tau) -$

$f(\varphi(t+\tau),t) + f(\varphi(t+\tau),t) - f(\varphi(t),t)$ and the first difference should be small because τ is an ϵ-translation number of f and the second because τ is an ϵ-translation number of φ. What then is involved? First the ϵ-translation number of f must be independent of the values of $\varphi(t+\tau)$ and in the second case f must be uniformly continuous in x, uniformly in t.

If on the other hand, we look at Definition 1.1, we would have to require that $\lim_{n\to\infty} f(\varphi(t+\alpha_n),t+\alpha_n)$ exists uniformly. Any proof of this would probably require that $\lim_{n\to\infty} f(x,t+\alpha_n)$ exists uniformly in (x,t).

2. Uniformly Almost Periodic Families. Using our principle that we use the definition of almost periodic appropriate to the situation, we introduce the proper notion in terms of $T(f,\epsilon)$.

Definition 2.1. A family $\mathfrak{F}$ of almost periodic functions is a uniformly almost periodic family if it is uniformly bounded, and if given $\epsilon > 0$, then $T(\mathfrak{F},\epsilon) = \bigcap_{f\in\mathfrak{F}} T(f,\epsilon)$ is relatively dense and includes an interval about 0.

3. Translation functions. In order to talk about uniformly almost periodic families (u.a.p. families) it is convenient to introduce Bochner's translation function. For a function $f \in AP(C)$ write $v_f(\tau) = \sup_{t\in R}|f(t+\tau) - f(t)|$. Then $T(f,\epsilon) \equiv \{\tau \mid v_f(\tau) < \epsilon\}$. The

properties of v_f are the following

(a) $v_f(\tau) \geq 0$; $v_f(-\tau) = v_f(\tau)$;

(b) $v_f(0) = 0$;

(c) $v_f(t+s) \leq v_f(t) + v_f(s)$;

(d) v_f is its own translation function

and

(e) f is in $AP(C)$ if and only if v_f is in $AP(C)$.

Properties (a) through (c) are easy to show. To prove (d) note that $v_f(t+s) - v_f(t) \leq v_f(s)$ so that $\sup_t |v_f(t+s) - v_f(t)| \leq v_f(s)$, but equality holds when $t = 0$. Now (e) follows immediately.

Note also that $\{\tau: v_f(\tau) < \epsilon\}$ contains an interval about 0 by the uniform continuity of functions in $AP(C)$. In fact this is equivalent to uniform continuity. Note also that properties (a) through (c) characterize translation functions. In fact, the proof of (d) uses only these facts.

<u>Theorem 2.2</u>. Suppose that for $f \in \mathfrak{F}$, $v(\tau) = \sup_{f\in\mathfrak{F}} v_f(\tau)$ is finite. Then v is a translation function. Consequently, the family $\mathfrak{F}$ is u.a.p. if and only if $v \in AP(C)$.

<u>Proof</u>: It is a straightforward calculation to show that v satisfies properties (a) through (c). If v is almost periodic, let $\tau \in T(v,\epsilon)$. Then $v(\tau) < \epsilon$ so that $v_f(\tau) < \epsilon$ for all $f \in \mathfrak{F}$ and hence $\tau \in T(f,\epsilon)$ for all $f \in \mathfrak{F}$ and the family is u.a.p. family. Conversely, suppose $\mathfrak{F}$ is a u.a.p. family. If $\tau \in T(\mathfrak{F},\epsilon)$ then

$v_f(\tau) \leq \epsilon$ for all $f \in \mathfrak{F}$ and so $v(\tau) \leq \epsilon$ and so $T(v,\epsilon)$ is relatively dense and contains an interval about zero so that v is uniformly continuous. To complete the proof one needs the finiteness of v at all points. This follows from the uniform boundedness of $\mathfrak{F}$ since $v_f(\tau) \leq 2\|f\|$.

In particular we may look at a finite set of functions in AP(C).

Corollary 2.3. Any finite set of functions in AP(C) is a u.a.p. family.

Proof: It is clear that $\sup_{f\in\mathfrak{F}} v_f$ with $\mathfrak{F}$ finite is bounded, and the family is uniformly bounded and equi-uniformly continuous. Furthermore, the sup of a finite family of a.p. functions is a.p., see 1.9.

In particular if f_1 and f_2 are in AP(C) then $T(f_1,\epsilon) \cap T(f_2,\epsilon)$ is relatively dense. This is a non-trivial fact.

4. Compactness in uniform convergence on compacta. The notion of u.a.p. family includes the notion of uniform boundedness and equicontinuity. In fact all u.a.p. families are uniformly bounded and equi-uniformly continuous, that is $\|f\| \leq M$ for all $f \in \mathfrak{F}$, where M depends only on the family, and there is a $\delta > 0$ so that $|f(x) - f(y)| < \epsilon$ for all $f \in \mathfrak{F}$ and $|x-y| < \delta$. Families with these two properties are pre-compact in the topology of uniform convergence on compact subsets of R. This fact will be used in almost every argument in our later sections on differential equations, so that there is a need for a precise statement.

Theorem 2.4. Let $\{f_n\}$ be a sequence of functions defined on R to C such that there is a number M for which $\|f_n\| \leq M$ for all n, and given $\epsilon > 0$, there is a $\delta > 0$ such that if $|x-y| < \delta$ then $|f_n(x) - f_n(y)| < \epsilon$ for all n. Then there is a subsequence which converges uniformly on every compact subset of R.

Proof: Let $\alpha_0 = \{n\}$, and $I_n = [-n,n]$. By the Arzela-Ascoli Theorem, the uniform boundedness and equi-uniform continuity imply that there is a sequence $\alpha_1 \subset \alpha_0$ so that $\{f_{(\alpha_1)_k}\}$ converges uniformly on I_1. Inductively, if α_n has been picked, then there is $\alpha_{n+1} \subset \alpha_n$ so that $\{f_{(\alpha_{n+1})_k}\}$ converges uniformly on I_{n+1}. Now let $\beta = \{(\alpha_1)_1, (\alpha_2)_2, \ldots\}$. Then $\beta \subset \alpha_n \subset \alpha_0$ for all n, so that $\{f_{\beta_n}\}$ converges uniformly on I_n.

The most common situation where this theorem applies is the case when $\|f_n\| \leq M$ and $\|f'_n\| \leq M$ for all n. In particular, if x is a bounded solution to a differential equation $x' = f(x,t)$ where f is bounded, then the set $\{x(t+a): a \in R\}$ is such a family. We will make repeated use of this fact.

5. Compactness in uniform norm.

We have seen that u.a.p. families have the property that from every sequence in the family one can extract a subsequence which converges uniformly on compact subsets. Our main object now, is to show that this extends to uniform convergence on R.

Theorem 2.5. If $\mathfrak{F}$ is a u.a.p. family, then from every sequence in $\mathfrak{F}$ one can extract a subsequence which converges uniformly on R.

Proof: This is a consequence of the fact that for such families, everything that happens on finite intervals almost happens on R. To be precise, let $\{f_n\}$ be a sequence in $\mathfrak{F}$, and note that we may take a subsequence that converges uniformly on compact subsets of R. Without changing notation assume this is again named $\{f_n\}$. If $\epsilon > 0$ is given, find an inclusion length $\ell = \ell(\epsilon/3)$ for $v(\tau) = \sup_{f\in\mathfrak{F}} v_f(\tau)$, and then a number N so that $\sup_{t\in[0,\ell]} |f_n(t) - f_m(t)| < \epsilon/3$ if $n,m \geq N$. Let x be a real number. There is a $\tau \in [-x, -x+\ell] \cap T(v,\epsilon/3)$. Then if $n,m \geq N$, we have

$|f_n(x) - f_m(x)| \leq |f_n(x) - f_n(x+\tau)| + |f_n(x+\tau) - f_m(x+\tau)| +$

$|f_m(x+\tau) - f_m(x)| < \epsilon$ since $\tau \in T(v, \epsilon/3)$ and $x+\tau \in [0,\ell]$. Now x is arbitrary so the sequence $\{f_n\}$ is a Cauchy sequence in $\| \ \|$.

It follows that if a u.a.p. family $\mathfrak{F}$ is closed in the uniform norm, then $\mathfrak{F}$ is compact. This is half of the next theorem.

Theorem 2.6. A family $\mathfrak{F} \subset AP(C)$ is compact if and only if it is closed in the uniform norm and u.a.p.

Proof: We need to show that a compact family is u.a.p. Let $\epsilon > 0$ be given, then there exists a finite set $f_1,..,f_N$ such that every $g \in \mathfrak{F}$ is within $\epsilon/3$ of one of $f_1,...,f_n$, i.e. $\|g - f_n\| < \epsilon/3$. It follows that $T(g,\epsilon) \supset T(f_n, \epsilon/3)$ by a straightforward triangle inequality argument. Thus $T(\mathfrak{F},\epsilon) \supset \bigcap_{n=1}^{N} T(f_n, \epsilon/3)$ and the latter is relatively dense by Corollary 2.3. It also contains an interval about 0. Finally, there is an M so that $\|f\| \leq M$ for all $f \in \mathfrak{F}$ since if $\|f_n\| > n$, then one gets for a subsequence, that $f_n \to f$ uniformly and hence f is unbounded but also is in $\mathfrak{F}$. This cannot be.

One notes that if one takes a given almost periodic translation function v and defines $\mathfrak{F} = \{f: f \in AP(C)$ and $v_f(\tau) \leq v(\tau)$ for all $\tau\}$ then $\mathfrak{F}$ is a u.a.p. family if and only if it is bounded. Note that $\mathfrak{F}$ is closed under additions of constants so that boundedness is not automatic. It is not automatic either that the functions are equi-uniformly continuous. For example, the functions $\{\sin nt \mid n = 1,2,\ldots\}$ are all periodic with one of the periods being 2π. Consequently $T = \bigcap_{n=1}^{\infty} T(\sin nt, \epsilon) \supset \{m2\pi: m = 0,\pm 1,\pm 2,\ldots\}$ so is relatively dense. The family is bounded, but not equi-uniformly continuous so that 0 is an isolated point of T for $\epsilon < 1$.

Note also that if $f_n \to f$ uniformly then $v_{f_n} \to v$. Gathering all this information together we can say that $\mathfrak{F} \equiv \{f: f \in AP(C)$, $\|f\| \leq M$ and $v_f(\tau) \leq v(\tau)$ for all $\tau\}$, where M is fixed and v is an almost periodic translation function, is a compact family.

6. <u>Bochner version of u.a.p.</u> We would like now to go to the Bochner version of u.a.p. since it will be more useful for differential equations.

<u>Theorem 2.7</u>. Let $\mathfrak{F}$ be a u.a.p. family. Given a sequence α',

there is $\alpha \subset \alpha'$ so that $T_\alpha f$ exists uniformly for all $f \in \mathfrak{F}$. In fact, the convergence is uniform in f, that is, given $\epsilon > 0$ there exists N so that $|f(t+\alpha_n) - f(t+\alpha_m)| < \epsilon$ for $n,m \geq M$, all $t \in R$ and all $f \in \mathfrak{F}$.

Proof: Let $v(\tau) = \sup_{f \in \mathfrak{F}} v_f(\tau)$. Then v is almost periodic so from a given α' we may find $\alpha \subset \alpha'$ so that $T_\alpha v$ exists uniformly. If $\epsilon > 0$ is given then $|v(\alpha_n) - v(\alpha_m)| < \epsilon$ if $n,m \geq N(\epsilon)$. Since $v(\alpha_n - \alpha_m) \leq |v(\alpha_n) - v(\alpha_m)|$ we have $v(\alpha_n - \alpha_m) < \epsilon$ and $v_f(\alpha_n - \alpha_m) < \epsilon$ for $f \in \mathfrak{F}$ and $n,m \geq N$. Thus $|f(t+\alpha_n - \alpha_m) - f(t)| < \epsilon$ for all t and by a change of variable $|f(t+\alpha_n) - f(t+\alpha_m)| < \epsilon$ for all t.

As a partial converse we have the following theorem.

Theorem 2.8. Let $\mathfrak{F}$ be a family of almost periodic functions which are uniformly bounded and from every sequence α' there is a $\alpha \subset \alpha'$ so that $T_\alpha f$ exists for every $f \in \mathfrak{F}$, the convergence being uniform in f. Then $\mathfrak{F}$ is a u.a.p. family.

Proof: Let $v(\tau) = \sup_{f \in \mathfrak{F}} v_f(\tau)$, and let α' be any sequence. We wish to show that there is $\alpha \subset \alpha'$ so that $T_\alpha v$ exists uniformly. To do this pick $\alpha \subset \alpha'$ so that $T_\alpha f$ exists uniformly in t and f. Now this means that $|f(t+\alpha_n) - f(t+\alpha_m)| < \epsilon/2$ if $n,m \geq N(\epsilon)$ independent of t and f. Rewriting this, we have

$$v_f(\alpha_n - \alpha_m) = \sup_t |f(t+\alpha_n - \alpha_m) - f(t)| \leq \epsilon/2 \quad \text{thus} \quad v(\alpha_n - \alpha_m) \leq \epsilon/2$$

if $n,m \geq N(\epsilon)$. But then $|v(t+\alpha_n - \alpha_m) - v(t)| \leq \epsilon/2$ for all t and from this one has $|v(t+\alpha_n) - v(t+\alpha_m)| \leq \epsilon/2$ for all t if $n,m \geq N(\epsilon)$. That is, $T_\alpha v$ exists uniformly. If v is continuous then v is almost periodic and we are done by Theorem 2.2. Now v is continuous if and only if it is continuous at 0. If $\alpha'_n \to 0$ and $v(\alpha'_n) \geq \epsilon > 0$ then pick $\alpha \subset \alpha'$ so that $T_\alpha f$ exists uniformly for $f \in \mathfrak{F}$. But then $T_\alpha f = f$ and we have $v_f(\alpha_n) \to 0$ uniformly in f. Thus $v(\alpha_n) \to 0$ and we have a contradiction.

7. Families indexed by E^n. Let E^n be either R^n or C^n with the usual Euclidean norm. We will henceforth consider functions of the form $f(x,t)$ where $x \in \Omega \subset E^n$ and $t \in R$ and f has values in E^n. The appropriate first assumption will be that for each $x \in \Omega$, the function $f(x,\cdot)$ is an almost periodic function. This is interpreted to mean that from every α' one can find $\alpha \subset \alpha'$ so that $T_\alpha f(x,\cdot)$ exists uniformly in the topology of E^n that is $\lim_{n\to\infty} f(x,t+\alpha_n)$ exists. We write all vectors as columns with f_i denoting the ith coordinate function.

Theorem 2.9. Let $f(x,t)$ be almost periodic in t for fixed x. Then each coordinate function $f_i(x,t)$ is almost periodic. Conversely, if $f_i(x,t)$ $i = 1,\ldots,n$ are almost periodic functions then the vector function $(f_1(x,t),\ldots,f_n(x,t))^T$ is an almost periodic function in E^n.

Proof: If $f(x,t)$ is almost periodic, then note that the projection onto the ith coordinate P_i, is uniformly continuous and hence

$P_i f(x,t)$ is clearly almost periodic. Conversely, the finite set $f_1,\dots,f_n$ is u.a.p. so if α' is given one has $\alpha \subset \alpha'$ so that $T_\alpha f_i(x,\cdot)$ exists uniformly in i and t. So $T_\alpha f(x,\cdot)$ exists uniformly in t in the E^n norm.

We have proved slightly more than we stated. The coordinate functions must be a u.a.p. family. Thus $T(f(x,\cdot),\epsilon)$ is always relatively dense and contains an interval about 0. In particular $v_f(\tau) = \sup_{t\in R}|f(x,t+\tau) - f(x,t)|$ is again a scalar almost periodic function, where $|\ |$ is now the E^n norm. All the previous theory for scalar valued function can be carried over verbatim, except of course, the lattice properties that make no sense here.

We began this Chapter with the problem of when compositions are in $AP(C)$ and we now transfer this problem to $AP(E^n)$ which is similarly defined. Since $f(\varphi(t),t)$ will require looking only at some bounded set of x values, looking at x in such sets is the appropriate assumption. One would also need some continuity in the function $f(x,t)$ as a function of two variables. The main result in this direction is

<u>Theorem 2.10.</u> Suppose $f(x,t) \in AP(E^n)$ for each $x \in K$ and is continuous on $K \times R$, where K is a compact subset of E^n. Then $f(x,t)$ is almost periodic in t uniformly for $x \in K$ if and only if f is continuous in x uniformly in t, that is, given $x \in K$ and $\epsilon > 0$, there is a $\delta(\epsilon,x)$ so that $y \in K$ and $|x-y| < \delta(\epsilon,x)$ implies that $|f(x,t) - f(y,t)| < \epsilon$ for all t.

Proof: If f has the continuity in x uniformly in t, then in fact it is uniformly continuous in x on the compact set K, so given $\epsilon > 0$ we let $\delta(\epsilon/3)$ be the modulus of uniform continuity for $\epsilon/3$. Then there are vectors $x_1,\ldots,x_n \in K$ so that for each $y \in K$, $|y-x_i| < \delta$ for some i. The family $f(x_i,\cdot)$ $i=1,\ldots,N$ is a u.a.p. family, so let $T = \bigcap_{i=1}^{n} T(f(x_i,\cdot),\epsilon)$. This is relatively dense. Let $\tau \in T$ and $y \in K$. Choose $|y-x_i| < \delta$ and then $|f(y,t+\tau)-f(y,t)| \leq |f(y,t+\tau)-f(x_i,t+\tau)| + |f(x_i,t+\tau)-f(x_i,t)| + |f(x_i,t)-f(y,t)| < \epsilon$. That is $T(f(K,\cdot),\epsilon) \supset T$ and is relatively dense and contains an interval about 0. Finally, $|f(y,t)| \leq |f(x_i,t)| + \epsilon \leq M$ so the family $\{f(x,\cdot)\,|\,x \in K\}$ is a u.a.p. family. Conversely, suppose this family is u.a.p. . Certainly $f(x,t)$ is uniformly continuous in x for $(x,t) \in K \times [0,L]$ for any finite L. So let $\epsilon > 0$ be given and let L = inclusion length for $T = T(f(K,\cdot),\epsilon/3)$. If $t \in R$, let $\tau \in [-t,-t+L] \cap T$. Let $\delta > 0$ be chosen so that $x,y \in K$ and $|x-y| < \delta$ imply $|f(x,t)-f(y,t)| \leq \epsilon/3$ for $t \in [0,L]$. Now for $|x-y| < \delta$ and $t \in R$ we have $|f(x,t)-f(y,t)| \leq |f(x,t)-f(x,t+\tau)| + |f(x,t+\tau)-f(y,t+\tau)| + |f(y,t+\tau)-f(y,t)| < \epsilon$ since $\tau \in T$ and $t+\tau \in [0,L]$. Thus f is uniformly continuous in x uniformly in t.

Some continuity in x is required for the above theorem since, if $f(x,t) = f_1(t)$ for $x \in A$ and $f(x,t) = f_2(t)$ for $x \in K-A$ where f_1 and f_2 are in $AP(E^n)$ and A is an arbitrary subset of K, then $f(x,t)$ is a u.a.p. family with no continuity in x.

8. Composition of AP functions. We are now ready to close the chapter with an answer to the question posed at the beginning.

Theorem 2.11. If $\varphi \in AP(E^n)$ and $f(x,\cdot) \in AP(E^n)$ uniformly for x in compact subsets of E^n, then $f(\varphi(t),t) \in AP(E^n)$.

Proof: Let K be a compact set containing the range of φ. If $\epsilon > 0$ is given, let $\tau \in T(\varphi,\delta) \cap T(f(K,\cdot),\epsilon/3)$ where $T(f(K,\cdot),\epsilon) = \bigcap_{x \in K} T(f(x,\cdot),\epsilon)$ is a relatively dense set as is its intersection with $T(\varphi,\delta)$, this being a u.a.p. family. The number δ is to be chosen so that $|f(x,t) - f(y,t)| < \epsilon/3$ when $|x-y| < \delta$. Then $|f(\varphi(t+\tau), t+\tau) - f(\varphi(t), t)| \leq |f(\varphi(t+\tau), t+\tau) - f(\varphi(t+\tau), t)| + |f(\varphi(t+\tau), t) - f(\varphi(t),t)| \leq \epsilon/3 + \epsilon/3 < \epsilon$ so $\tau \in T(f(\varphi(t), t), \epsilon)$ and this latter set is relatively dense.

We shall have more to say about this composition after we have done the harmonic analysis of almost periodic functions.

9. Notes. Our definition of uniformly almost periodic family is slightly different from the usual one in that we require uniform boundedness and equicontinuity in the definition. The usual term, found in Besicovitch's book, for example, is homogeneous family.

The notion of translation function is another innovation of Bochner, Beiträge zur theorie der fastperiodischen Funktionen, I. Funktionen einer Variablen., Math Ann. 96(1927), 119-147.

The results on functions indexed by E^n are a part of the folklore of the subject. Pieces can be found in various sources.

An early reference to some of these ideas is in Bochner and von Neumann, Trans. Amer. Math. Soc. (1935), or Amerio-Prouse [30] where almost periodic functions with values in Banach spaces are considered.

<u>Chapter 3</u>

The Fourier Series Theory

1. <u>Introduction</u>. We should like to recall some of the main aspects of the fourier series theory of periodic functions. Let f be periodic with period T, then there is the formal relation

$$f \sim \sum_{n=-\infty}^{\infty} a_n e^{i\lambda_n t} \quad \text{where} \quad \lambda_n = \frac{2n\pi}{T} \quad \text{and}$$

$$a_n = \frac{1}{T}\int_0^T f(t) e^{-\lambda_n t} dt . \qquad (3.1)$$

Questions of convergence aside, the main results are:

1) the mapping from f to the numbers $\{(a_n, \lambda_n)\}$ is one to one;

2) Parseval's equation holds, that is

$$\frac{1}{T}\int_0^T |f(t)|^2 dt = \sum_{n=-\infty}^{\infty} |a_n|^2 ;$$

and

3) there are altered partial sums which approximate f; that is, there are numbers $b_{n,m}$ so that for fixed m, $b_{n,m} = 0$ for large n and $|f(t) - \sum_{n=-\infty}^{\infty} b_{nm} a_n e^{i\lambda_n t}| < 1/m$ for all t.

To extend the above results to almost periodic functions we need to replace λ_n by a more general real number and replace the formula 3.1 by something more appropriate. No fixed T will give a satisfactory theory since many almost periodic functions would have the same Fourier series. On the other hand, the Fourier transform would not exist in general. A compromise, which is the correct one, is take an "average" of the Fourier transform, that is, take limits as $T \to \infty$ in 3.1.

We will develop the Fourier series theory for functions in $AP(E^n)$ with parameters in E^n.

2. Mean values. We investigate first the Bohr transform of an almost periodic function,

$$a(f,\lambda,x) = \lim_{T\to\infty} \frac{1}{T}\int_0^T f(x,t)e^{-i\lambda t}\,dt \tag{3.2}$$

for $f(x,\cdot) \in AP(E^n)$ uniformly for x in compact subsets of E^n. We will not repeat this hypothesis explicitly in the discussion. Note that for each fixed (λ,f,x), $a(f,\lambda,x) \in E^n$.

Theorem 3.1. For any $\lambda \in R$, $a(\lambda,f,x)$ exists uniformly for x in compact sets and is continuous in x.

Proof: Let K be a compact set in E^n, and $\lambda = 0$. Let M be a bound for $\|f(x,\cdot)\|$ over $x \in K$, and $T(K,\epsilon)$ be the ϵ-translation sets for $f(x,\cdot)$ as x ranges over K. If $\epsilon > 0$ is given, let ℓ be the inclusion length for $T(K,\epsilon/8)$ and S a real number so that $S > \frac{16M\ell}{\epsilon}$. Then

$$\frac{1}{nS}\int_0^{nS} f(x,t)\,dt = \sum_{k=0}^{n-1} \frac{1}{nS}\int_{kS}^{(k+1)S} f(x,t)\,dt .$$

Write $\int_{kS}^{(k+1)S} f(x,t)\,dt = \int_{kS-\tau}^{(k+1)S-\tau} f(x,t+\tau)\,dt$

$= \int_0^S f(x,t)\,dt + \int_0^S (f(x,t+\tau) - f(x,t))\,dt + \int_{kS-\tau}^{0} f(x,t+\tau)\,dt +$

$\int_S^{(k+1)S-\tau} f(x,t+\tau)\,dt$. If $\tau \in T(K,\epsilon/8) \cap (kS,kS+\ell)$ then we have the

estimate $\left|\int_{kS}^{(k+1)S} f(x,t)\,dt - \int_0^S f(x,t)\,dt\right| \le \frac{S\epsilon}{8} + 2M\ell$ and

consequently, $\left|\frac{1}{nS}\int_0^{nS} f(x,t)\,dt - \frac{1}{S}\int_0^S f(x,t)\,dt\right| \le \epsilon/8 + \frac{2M\ell}{S} < \epsilon/4.$

Now if T is a real number, there is a unique n so that

$nS \le T \le (n+1)S$ and thus (f is bounded)

$\lim_{T\to\infty} \left|\frac{1}{T}\int_0^T f(x,t)\,dt - \frac{1}{nS}\int_0^{nS} f(x,t)\,dt\right| = 0$ implies that

$$\left|\frac{1}{T}\int_0^T f(x,t)\,dt - \frac{1}{nS}\int_0^{nS} f(x,t)\,dt\right| < \epsilon/4 \qquad (3.3)$$

if $T \ge T_0$. Note T_0 does not depend on x but on M.

Now if T_1 and $T_2 \ge T_0$ and n_1 and n_2 are chosen so that $n_iS \le T_i \le (n_i+1)S$ then $\left|\frac{1}{T_1}\int_0^{T_1} f(x,t)\,dt - \frac{1}{T_2}\int_0^{T_2} f(x,t)\,dt\right| \le$

$\left|\frac{1}{T_1}\int_0^{T_1} f(x,t)\,dt - \frac{1}{n_1S}\int_0^{n_1S} f(x,t)\,dt\right| + \left|\frac{1}{n_1S}\int_0^{n_1S} f(x,t)\,dt - \frac{1}{S}\int_0^S f(x,t)\,dt\right|$

$+ \left|\frac{1}{S}\int_0^S f(x,t)\,dt - \frac{1}{n_2S}\int_0^{n_2S} f(x,t)\,dt\right|$

$+ \left|\frac{1}{n_2S}\int_0^{n_2S} f(x,t)\,dt - \frac{1}{T_2}\int_0^{T_2} f(x,t)\,dt\right| < \epsilon$, so the

$\lim_{T\to\infty} \frac{1}{T}\int_0^T f(x,t)\,dt = a(f,0,x)$ exists uniformly for $x \in K$. Noting that $\frac{1}{T}\int_0^T f(x,t)\,dt$ is continuous in x, it follows that $a(f,0,x)$ is (uniformly) continuous on K. To do the case $a(f,\lambda,x)$ note that $a(f,\lambda,x) = a(e^{-i\lambda t}f,0,x)$ and this latter exists since $e^{-i\lambda t}f(x,t) \in AP(E^n)$. We use this reduction in the next Corollary.

Corollary 3.2. For $f \in AP(E^n)$ $a(f,\lambda,x) = \lim_{T\to\infty} \frac{1}{T}\int_a^{a+T} f(x,t)e^{-i\lambda t}dt$ where the limit is uniform for $a \in R$.

Proof: We look at equation 3.3 and let $n \to \infty$ to get the estimate

$$\left|\frac{1}{T}\int_0^T f(x,t)\,dt - a(f,0,x)\right| < \epsilon/4 \qquad (3.4)$$

where now the estimate for T depends only on the numbers ϵ, M, and ℓ. Consequently, this limit exists uniformly for any class of functions for which M and ℓ can be picked independent of the class, i.e., any u.a.p. family. If $f(x,t)$ is in such a family, so is $f(x,t+a)$ for any $a \in R$. But $\frac{1}{T}\int_0^T f(x,t+a)\,dt = \frac{1}{T}\int_a^{a+T} f(x,t)\,dt$. In (3.4) we make this replacement, as well as the fact that $a(f(x,t+a),0,x) = a(f(x,t),0,x)$ to get the Corollary. The last identity is clear since

$$\frac{1}{T}\int_a^{a+T} f(x,t)\,dt = \frac{1}{T}\int_0^T f(x,t)\,dt + \frac{1}{T}\left(-\int_0^a f(x,t)\,dt + \int_T^{a+T} f(x,t)\,dt\right)$$

and the parenthesis is bounded by $2aM$.

We now can make the formal correspondence between an almost periodic function and its fourier series $f(x,t) \sim \sum a(f,\lambda,x)e^{i\lambda t}$

where the coefficient function is given by 3.2.

3. Bessel's Inequality. As a first step in getting Parseval's equation we get one of the inequalities and some of the properties of the Bohr transform. For example, it is clear that the Bohr transform is linear in f. Thus the one to one property is simply the fact, that the Bohr transform is zero only for the zero function. As a first step we note that $a(e^{i\lambda t}1,0) = 0$ if $\lambda \neq 0$ and $a(1,0) = 0$ where 1 is the constant vector function, 1 in each component. The number $a(f,0,x)$ is called the mean value of f, which for convenience we denote as $M_t(f(x,t))$ in the following proof.

Theorem 3.3. For any finite set of distinct real numbers $\lambda_1,\ldots,\lambda_N$, we have $\sum_{n=1}^{N} |a(f,\lambda_n,x)|^2 \leq M_t\{|f(x,t)|^2\}$.

Proof: Note that $|f(x,t)|^2 = \langle f(x,t), \overline{f(x,t)}\rangle$, where $\langle\ ,\ \rangle$ denotes the usual inner product in E^n, is in $AP(R)$ so has a mean value. Thus $M_t(|f(x,t) - \sum_{n=1}^{N} a(f,\lambda_n,x)e^{i\lambda_n t}|^2)$

$$= M_t(\langle f(x,t) - \sum_{n=1}^{N} a(f,\lambda_n,x)e^{i\lambda_n t},\ \overline{f(x,t)} - \sum_{n=1}^{N} \overline{a(f,\lambda_n,x)e}^{-i\lambda_n t}\rangle)$$

$$= M_t(|f(x,t)|^2) - \sum_{n=1}^{N} \langle\overline{a(f,\lambda_n,x)},\ M_t(f(x,t)e^{-i\lambda_n t})\rangle$$

$$- \sum_{n=1}^{N} \langle a(f,\lambda_n,x),\ M_t(\overline{f(x,t)e}^{i\lambda_n t})\rangle + \sum_{k=1}^{N}\sum_{j=1}^{N} |\overline{a(f,\lambda_k,x)}|^2 M(e^{i(\lambda_k-\lambda_j)t})$$

$$= M_t(|f(x,t)|^2) - \sum_{n=1}^{N} |a(f,\lambda_n,x)|^2 - \sum_{n=1}^{N} |a(f,\lambda_n,x)|^2 + \sum_{n=1}^{N} |a(f,\lambda_n,x)|^2$$

since $M_t(f(x,t)e^{-i\lambda_n t}) = a(f,\lambda_n,x)$ and $M(e^{i(\lambda_k-\lambda_j)t}) = 0$ if $\lambda_k \neq \lambda_j$ and 1 if $\lambda_k = \lambda_j$. Thus we have

$$0 \leq M_t(|f(x,t) - \sum_{n=1}^{N} a(f,\lambda_n,x)e^{i\lambda_n t}|^2)$$

$$= M_t(|f(x,t)|^2) - \sum_{n=1}^{N} |a(f,\lambda_n,x)|^2 . \tag{3.5}$$

This completes the proof.

Since for x in a compact set $|f(x,t)|^2$ is uniformly bounded, the set of λ such that $|a(f,\lambda,x)| \geq \frac{1}{n}$ must be finite or else Theorem 3.3 would be violated. It follows that the set of λ for which $a(f,\lambda,x) \neq 0$ must be countable. We need a slightly stronger result. The above holds for any fixed x, but we want to show that as functions of $x \in K$, at most at countable number of $a(f,\lambda,x)$ are non-zero. That is, there are a countable number of λ's so that if μ is not in this set there $a(f,\mu,x) \equiv 0$ for $x \in K$. To do this, we look at Theorem 3.3 and take supremum over $x \in K$ for K compact, we then get $\sum_{n=1}^{N} \sup_{x\in K}|a(f,\lambda_n,x)|^2 \leq M^2$ where M is a uniform bound for $|f(x,t)|^2$. Now the above argument applies.

<u>Corollary 3.4</u>. If $f(x,t) \in AP(E^n)$ uniformly for x in compact sets, then there is a countable set of real numbers Λ such that $a(f,\lambda,x) \equiv 0$ on K if $\lambda \notin \Lambda$.

<u>Definition 3.5</u>. The set Λ for which $a(f,\lambda,x) \not\equiv 0$ is called the

set of exponents of f on K. The numbers $a(f,\Lambda,x)$ for $\lambda \in \Lambda$ are the fourier coefficients. Denote Λ by $\exp(f)$.

Since Λ is countable it can be enumerated by the positive integers and one can write $f(x,t) \sim \sum a(f,\lambda_n,x)e^{i\lambda_n t}$.

Of course no convergence is implied by this representation. We have though, that since Theorem 3.3 holds for any finite set of numbers in Λ, that the sum over Λ converges, i.e. Bessel's inequality holds.

Theorem 3.6. (Bessel's inequality) For $f(x,t) \in AP(E^n)$

$$\sum_{n=1}^{\infty} |a(f,\lambda_n,t)|^2 \leq M_t(|f(x,t)|^2)$$

One other comment worthy of note is the fact that Λ is invariant under translation of functions, that is, $f(x,t+a) \sim \sum a(f,\lambda,x)e^{i\lambda_n a}e^{i\lambda_n t}$ which can be verified by direct calculation. Also a uniformly convergent series of the form $\sum_{n=1}^{\infty} a_n e^{i\lambda_n t}$ is its own fourier series. Indeed

$$\lim_{T\to\infty}\frac{1}{T}\int_0^T \Big(\sum_{n=1}^{\infty} a_n e^{i\lambda_n t} e^{-i\mu t}\Big) = \sum_{n=1}^{\infty} a_n\Big(\lim_{T\to\infty}\frac{1}{T}\int_0^T e^{i(\lambda_n-\mu)t}\,dt\Big) = 0 \text{ if}$$

$\mu \neq \lambda_n$ for all n and $= a_n$ if $\mu = \lambda_n$.

The Bohr transform also exhibits some continuity as a function of f.

Theorem 3.7. For any λ and $f,g \in AP(E^n)$ one has

$|a(f,\lambda,x) - a(g,\lambda,x)| \leq \|f(x,\cdot) - g(x,\cdot)\|$. Consequently, if $\{f_n(x,t)\}$ converges uniformly to $f(x,t)$ on $K \times R$, then $a(f_n,\lambda,x) \to a(f,\lambda,x)$ uniformly on K . In particular, the fourier series of f_n converges formally to the fourier series of f .

Proof: Just apply the definition of $a(f,\lambda,x)$.

One final result for scalar valued functions.

Theorem 3.8. If f is a non-negative almost periodic function, $f \not\equiv 0$, then $a(f,0) > 0$.

Proof: Let $f(x_0) = M > 0$ and pick $\delta > 0$ so that $f(x) \geq \frac{2M}{3}$ on $(x_0-\delta,x_0+\delta)$. Let ℓ be an inclusion length for $T(f,M/3)$ and take $\ell > 2\delta$. If a is a real number, find $\tau \in T(f,M/3) \cap [a+\delta-x_0,a+\delta-x_0+\ell]$. Then $x_0-\delta+\tau \in [a,a+\ell]$. Either $x_0+\tau$ or $x_0-2\delta+\tau \in [a,a+\ell]$ since $\ell > 2\delta$. In the first case if $t \in (x_0-\delta+\tau,x_0+\tau)$ then $|f(t)| \geq |f(t+\tau)| - |f(t+\tau) - f(t)|$ $\geq \frac{2M}{3} - \frac{M}{3} = \frac{M}{3}$. The second case is handled similarly. In either case $\int_a^{a+\ell} f(t)dt > \frac{M}{3}\delta$ since on a subinterval of length δ , $f(t) \geq M/3$. Now write $a = (n-1)\ell$ to get $\int_{(n-1)\ell}^{n\ell} f(t)dt \geq \frac{M}{3}\delta$ and a fortiori

$$\frac{1}{N\ell}\int_0^{N\ell} f(t)dt = \frac{1}{N\ell}\sum_{n=1}^{N}\int_{(n-1)\ell}^{n\ell} f(t)dt \geq \frac{M\delta}{3\ell} .$$

Letting $N \to \infty$ we get

$$a(f,0) \geq \frac{M\delta}{3\ell} .$$

4. Mean Convergence. In analogy with the L_1 theory of convergence, we can look at convergence of sequences in the mean M .

Definition 3.9. A sequence of functions $f_n \in AP(E^n)$ converges in the mean if $M(|f_n - f_m|^2) \to 0$ as $n,m \to \infty$.

It is easy to see that if $\|f_n - f_m\| \to 0$ as $n,m \to \infty$, that is $\{f_n\}$ converges uniformly, then the sequence converges in the mean. In fact $M(|f_n - f_m|^2) \leq \|f_n - f_m\|^2$. The converse is not in general true, even for periodic functions. However, for u.a.p. families it is.

Theorem 3.10. If $f_n(x,t) \in AP(E^n)$ and if the family $\{f_n(x,t)$: n is a positive integer and $x \in K\}$ is a u.a.p. family then convergence in the mean is equivalent to uniform convergence.

Proof: It is sufficient to show that uniform convergence follows from mean convergence. To do this, we prove a stronger version of the contrapositive. Indeed, we show that for every $\epsilon > 0$, there is a $\rho > 0$ so that if $\|f_n(x,t) - f_m(x,t)\| > \epsilon$ for some $x \in K$, then $M_t(|f_n(x,t) - f_m(x,t)|^2) > \rho$. To prove this let $\epsilon > 0$ be given and $(-\delta,\delta) \subset T(v,\epsilon/6)$ where v is the translation function for all the f_n as x ranges over the compact set K. If $\|f_n(x,t) - f_m(x,t)\| > \epsilon$ for some $x \in K$, let $|f_n(x,t_0) - f_m(x,t_0)| > \epsilon$. Then on $(t_0-\delta, t_0+\delta)$ $|f_n(x,t) - f_m(x,t)| > \epsilon/2$. If ℓ is the inclusion length for $T(v,\epsilon/6)$ and $\ell > 2\delta$ then every interval $((n-1)\ell, n\ell)$ contains one half of an interval $(t_0+\tau-\delta, t_0+\tau+\delta)$ where $\tau \in T(v,\epsilon/6)$. This is a repeat of the argument of Theorem 3.8. In this half interval $|f_n(x,t) - f_m(x,t)| > \epsilon/6$. Hence

$$\int_{(n-1)\ell}^{n\ell} |f_n(x,t) - f_m(x,t)|^2 dt > (\tfrac{\epsilon}{6})^2 \delta \quad \text{so}$$

$$M_t(|f_n(x,t) - f_m(x,t)|^2) = \lim_{N\to\infty} \frac{1}{n\ell}\int_0^{n\ell} |f_n(x,t) - f_m(x,t)|^2 dt \geq (\tfrac{\epsilon}{6})^2 \frac{\delta}{\ell}$$

as in the proof of Theorem 3.8. Take $\rho = (\frac{\epsilon}{6})^2 \frac{\delta}{\ell}$.

5. Parseval's equation. We want to now show that the Bessel inequality is in fact an equality. In order to do this, we make several reductions. First, it is sufficient to prove Parseval's equation for the case when $a(f,0,x) \equiv 0$. Indeed if f is given then $f - a(f,0,x)$ has mean value zero and the same Bohr transform for $\lambda \neq 0$. A glance at 3.5 then shows that

$$M_t(|f(x,t)|^2) = M_t(|f(x,t) - a(f,0,x)|^2) + |a(f,0,x)|^2$$

$$= \sum_{\lambda_n \neq 0} |a(f,\lambda_n,x)|^2 + |a(f,0,x)|^2.$$

The second reduction is that for uniform convergence it is sufficient to show that Parseval's equation holds pointwise for $x \in K$. In fact the partial sums $\sum_{n=1}^{N} |a(f,\lambda_n,x)|^2 = S_n(x)$ are continuous functions and $S_{n+1} \geq S_n$ on K. By Dini's Theorem, pointwise convergence to the continuous function $M_t(|f(x,t)|^2)$ is then equivalent to uniform convergence. Thus we may restrict ourselves to a fixed x and so we drop it from our notation and consider a single function $f(t) \in AP(E^n)$.

Finally, in view of Bessel's inequality it is sufficient to

show that given $\epsilon > 0$, one can find N so that

$$\sum_{n=1}^{N} |a(f,\lambda_n)|^2 \geq M(|f|^2) - \epsilon.$$

The main tool in our proof of this theorem is the Parseval equation for Fourier transforms of functions with compact support. Specifically, let F be continuous on an finite interval I and 0 outside I. Then for $A(\lambda) = \int_{-\infty}^{\infty} F(x)e^{-i\lambda x}dx$ one has

$$\int_{-\infty}^{\infty} |F(x)|^2 dx = \frac{1}{2\pi}\int_{-\infty}^{\infty} |A(\lambda)|^2 d\lambda. \qquad (3.6)$$

For any $T > 0$ let $f_T(x) = f(x)$ on $[0,T]$ and $f_T(x) = 0$ elsewhere. Then $a_T(\lambda) = \frac{1}{T}\int_{-\infty}^{\infty} f_T(x)e^{-i\lambda x}dx = \frac{1}{T}\int_0^T f(x)e^{-i\lambda x}dx$, and

$\frac{1}{T}\int_0^T |f(x)|^2 = \frac{1}{T}\int_{-\infty}^{\infty} |f_T(x)|^2 dx = \frac{T}{2\pi}\int_{-\infty}^{\infty} |a_T(\lambda)|^2 d\lambda$. To prove Parseval's equation we thus need to show that for every $\epsilon > 0$, there is a $T_0 > 0$ so that if $T > T_0$

$$\frac{T}{2\pi}\int_{-\infty}^{\infty} |a_T(\lambda)|^2 d\lambda < \sum |a(\lambda)|^2 + \epsilon \qquad (3.7)$$

where we write $a(\lambda)$ for $a(f,\lambda)$. We are assuming that $a(0) = 0$.

We will show inequality 3.7 by a sequence of lemmas.

<u>Lemma 3.11</u>. For every λ_0 and $\epsilon > 0$, there are numbers $\delta > 0$ and $T_0 > 0$ so that

$$\frac{T}{2\pi}\int_{\lambda_0-\delta}^{\lambda_0+\delta} |a_T(\lambda)|^2 d\lambda < |a(\lambda_0)|^2 + \epsilon \quad \text{when} \quad T > T_0.$$

<u>Proof</u>: Replacing f by $f(t)e^{-i\lambda_0 t}$, replaces $a_T(\lambda)$ by $a_T(\lambda_0+\lambda)$ so we may assume that $\lambda_0 = 0$. Let $F(x) = \frac{1}{a}\int_x^{x+a} f_T(y)dy$. Then noting that F is zero except on a finite interval,

$$\frac{1}{T}\int_{-\infty}^{\infty} F(x)e^{-i\lambda x}dx = \frac{1}{aT}\int_{-\infty}^{\infty}\left(\int_x^{x+a} f_T(y)dy\right)e^{-i\lambda x}dx$$

$$= \frac{1}{aT}\int_{-\infty}^{\infty}\left(\int_y^{y+a} e^{-i\lambda x}dx\right) f_T(y)dy = \frac{1}{T}\int_{-\infty}^{\infty} e^{i\lambda y}f_T(y)\ \frac{e^{-i\lambda a}-1}{i\lambda a}\ dy$$

$= a_T(\lambda)\ \dfrac{e^{i\lambda a-1}}{i\lambda a}$. By Parseval's equation,

$$\frac{1}{T}\int_{-\infty}^{\infty} |F(x)|^2 dx = \frac{T}{2\pi}\int_{-\infty}^{\infty} |a_T(\lambda)|^2 \left(\frac{e^{i\lambda a-1}}{i\lambda a}\right|^2 d\lambda. \qquad (3.8)$$

If $\epsilon > 0$ is given choose a so that $\frac{1}{a}\int_x^{x+a} f(t)dt < \epsilon$ uniformly in x. Pick $T > a$. If $x \in [0, T-a]$ then $\frac{1}{a}\int_x^{x+a} f(t)dt = F(x)$ so $|F(x)| < \epsilon$ there. If $x \notin (-a,T)$ then $F(x) = 0$. Elsewhere $|F(x)| < \|f\|$. Thus

$$\frac{1}{T}\int_{-\infty}^{\infty} |F(x)|^2 dx \le \epsilon^2 + \frac{2a\|f\|^2}{T} .$$

Also for a fixed a we find $\delta > 0$ so that $\left|\frac{e^{i\lambda a}-1}{i\lambda a}\right| \ge 1/2$, $|\lambda| < \delta$.

Combining these last two estimates into 3.8, we get

$$\frac{T}{2\pi}\int_{-\delta}^{\delta} |a_T(\lambda)|^2 d\lambda \le \frac{4T}{2\pi}\int_{-\delta}^{\delta} |a_T(\lambda)|^2 \left|\frac{e^{i\lambda a}-1}{i\lambda a}\right|^2 d\lambda$$

$$\leq \frac{4T}{2\pi}\int_{-\infty}^{\infty} |a_T(\lambda)|^2 \left|\frac{e^{i\lambda a}-1}{i\lambda a}\right|^2 d\lambda \leq \frac{2}{T}\int_{-\infty}^{\infty} |F(x)|^2 dx \leq 2(\epsilon^2 + \frac{2a\|f\|^2}{T}) \quad \text{if}$$

$T > a$. Taking T large gives the estimate for the Lemma.

This says roughly that the Bohr transform is continuous in λ and that $a_T(\lambda)$ uniformly approximates $a(\lambda)$ in small intervals. The content of the next lemmas is that $a_T(\lambda)$ is concentrated on a finite union of small intervals.

<u>Lemma 3.12</u>. Let $v(t)$ be the translation function of an almost periodic function. Let $S(\epsilon) = \{\lambda \mid |e^{i\lambda t}-1| \leq 1$ for all $t \in T(v,\epsilon)\}$. Then $S(\epsilon)$ is finite. Furthermore, if $S(\epsilon) = \{\lambda_1,\ldots,\lambda_n\}$ then for every $\rho > 0$ there is an $A > 0$ so that if $|\lambda-\lambda_i| \geq \rho$ then $|e^{i\lambda t}-1| > 1$ for some $t \in [0,A] \cap T(v,\epsilon)$.

<u>Proof</u>: To show that $S(\epsilon)$ is finite we note that $T(v,\epsilon)$ contains an interval $(-\delta,\delta)$. (May assume $\delta < 1$.) If $|\lambda| > \frac{\pi}{3\delta}$, then there is a number in this interval for which $|e^{i\lambda t}-1| > 1$ since $\lambda(-\delta,\delta)$ covers the interval $(-\pi/3,\pi/3)$. Thus $S(\epsilon) \subset (\frac{-\pi}{3\delta},\frac{\pi}{3\delta})$. If $\lambda \in S(\epsilon)$ then $|\lambda t| < \frac{\pi}{3} \bmod (2\pi)$ so if λ_1 and λ_2 are in $S(\epsilon)$ then $|(\lambda_1-\lambda_2)t| < \frac{2\pi}{3} \bmod (2\pi)$ so that the interval $(\frac{2\pi}{3}|\lambda_1-\lambda_2|^{-1}, \frac{4\pi}{3}|\lambda_1-\lambda_2|^{-1}) \cap T(v,\epsilon) = \emptyset$. This interval has length $\frac{2\pi}{3}|\lambda_1-\lambda_2|^{-1}$ and therefore must be smaller than ℓ, the inclusion interval for $T(v,\epsilon)$. Hence $|\lambda_1-\lambda_2| \geq \frac{2\pi}{3\ell}$ and $S(\epsilon)$ is finite.

If $\rho > 0$ is given, consider the compact set $\{\lambda \mid |\lambda-\lambda_i| \geq \rho \ i=1,\ldots,N, \ |\lambda| \leq \frac{\pi}{3}\}$. This set is covered by a union of the open sets $\{\lambda \mid |e^{it\lambda}-1| > 1\}$ as t ranges over

$T(v,\epsilon) \cap [0,\infty)$. We select a finite subcover indexed by $t_1,\ldots,t_m$. Let A be greater than δ and all t_i. This completes the proof.

Lemma 3.13. Let $\varphi(t) = \max(0, \epsilon - v(t))$ for v an a.p. translation function and ϵ a positive number. Then there exist $\lambda_1,\ldots,\lambda_N$ such that for every $\rho > 0$ there is a $S > 0$ so that

$$\frac{1}{S}\int_0^S |e^{i\lambda t}-1|^2 \varphi(t)\,dt > \frac{1}{18}\,\frac{1}{S}\int_0^S \varphi(t)\,dt$$

when $|\lambda - \lambda_i| \geq \rho \quad i = 1,\ldots,N.$

Proof: The function φ is almost periodic by Theorem 1.9 and clearly non-negative but not identically zero. By Theorem 3.8, its mean value is positive, say $M(f) = M$. In Lemma 3.12, let $\epsilon = M/2$ to get numbers $\lambda_1,\ldots,\lambda_N$ so that when $\rho > 0$ is chosen there is an $A > 0$ given by that lemma. Let $S > 2A$, and let $|\lambda-\lambda_i| \geq \rho$ and $\tau \in (0,A) \cap T(v,\epsilon)$ so that $|e^{i\lambda\tau}-1| > 1$. Then

$$\frac{1}{S}\int_0^S |e^{i\lambda t}-1|^2\varphi(t)\,dt \geq \frac{1}{2S}\int_0^{S-A} |e^{i\lambda t}-1|^2\varphi(t)\,dt$$

$$+ \frac{1}{2S}\int_0^S |e^{i\lambda t}-1|^2\varphi(t)\,dt \geq \frac{1}{2S}\int_0^{S-A} |e^{i\lambda t}-1|^2\varphi(t)\,dt + \frac{1}{2S}\int_\tau^{\tau+S-A} |e^{i\lambda t}-1|^2\varphi(t)\,dt$$

$$= \frac{1}{2S}\int_0^{S-A} |e^{i\lambda t}-1|^2\varphi(t)\,dt + \frac{1}{2S}\int_0^{S-A} |e^{i\lambda(t+\tau)}-1|^2\varphi(t+\tau)\,dt$$

$$= \frac{1}{2S}\int_0^{S-A} \left(|e^{i\lambda t}-1|^2\varphi(t) + |e^{i\lambda(t+\tau)}-1|^2\varphi(t+\tau)\right)dt.$$ Since $|\lambda\tau| > \pi/3 \bmod 2\pi$ it is not true that both $|\lambda t| \leq \pi/6 \bmod 2\pi$ and $|\lambda(t+\tau)| \leq \pi/6 \bmod 2\pi$. Thus $\max\{|e^{i\lambda t}-1|^2, |e^{i\lambda(t+\tau)}-1|^2\}$

$> |e^{i\pi/6}-1|^2 > 1/4$. Also since $\tau \in T(v,\epsilon)$, $v(t+\tau) \leq v(t) + v(\tau)$ $\leq v(t) + \epsilon$ so that $\varphi(t+\tau) \geq \varphi(t) - \epsilon$. Consequently,

$$\frac{1}{2S}\int_0^{S-A} (|e^{i\lambda t}-1|^2\varphi(t) + |e^{i\lambda(t+\tau)}-1|^2\varphi(t+\tau))dt$$

$$\geq \frac{1}{2S}\int_0^{S-A} (|e^{i\lambda t}-1|^2 + |e^{i\lambda(t+\tau)}-1|^2)(\varphi(t) - \epsilon)dt \geq \frac{1}{8S}\int_0^{S-A} [\varphi(t)-\epsilon]dt .$$

Thus for $S > 2A$, $\frac{1}{S}\int_0^S |e^{i\lambda t}-1|^2\varphi(t)dt > \frac{1}{8S}\int_0^{S-A} [\varphi(t)-\epsilon]dt$. As $S \to \infty$ the right hand side has limit $\frac{1}{8}(M-\epsilon) = \frac{M}{16}$ so eventually is larger than $\frac{1}{18S}\int_0^S \varphi(t)dt$ which has limit $\frac{M}{18}$.

<u>Lemma 3.14</u>. For every $\epsilon > 0$ there is a finite set of numbers $\lambda_1,\ldots,\lambda_N$ so that for every $\rho > 0$ and $T > T_0(\rho)$

$$\frac{T}{2\pi}\int_u |a_T(\lambda)|^2 d\lambda < \epsilon$$

where $U = \{\lambda \mid |\lambda-\lambda_i| \geq \rho \ i=1,\ldots,N\}$.

<u>Proof</u>: Write $\psi(x) = f_T(x+\tau) - f_T(x)$ so that

$$\frac{1}{T}\int_{-\infty}^{\infty} \psi(x)e^{-i\lambda x}dx = a_T(\lambda)(e^{i\lambda\tau}-1) \quad \text{and hence}$$

$$\frac{1}{T}\int_{-\infty}^{\infty} |\psi(x)|^2dx = \frac{T}{2\pi}\int_{-\infty}^{\infty} |a_T(\lambda)|^2|e^{i\lambda\tau}-1|^2d\lambda.$$ Note that $|\psi| \leq v(\tau)$ if $x, x+\tau \in [0,T]$; $\psi = 0$ if both are not in $[0,T]$ and otherwise $|\psi| \leq \|f\|$. Thus $\frac{1}{T}\int_{-\infty}^{\infty} |\psi(x)|^2dx \leq v(\tau)^2 + \frac{2\tau\|f\|^2}{T}$. If $T > \frac{2\tau\|f\|^2}{\epsilon^2}$ and $v(\tau) \leq \epsilon$ then $\frac{T}{2\pi}\int_{-\infty}^{\infty} |a_T(\lambda)|^2|e^{i\lambda\tau}-1|^2d\lambda \leq 2\epsilon^2$.

Now let $\varphi(t) = \max(\epsilon - v(t), 0)$. Then

$$\frac{T}{2\pi}\int_{-\infty}^{\infty} |a_T(\lambda)|^2 |e^{i\lambda\tau}-1|^2 \varphi(\tau)\,d\lambda \le 2\epsilon^2\varphi(\tau)$$

for all τ. Thus

$$\frac{T}{2\pi}\int_{-\infty}^{\infty} |a_T(\lambda)|^2 \frac{1}{S}\int_0^S |e^{i\lambda\tau}-1|^2\varphi(\tau)\,d\tau\,d\lambda \le 2\epsilon^2 \frac{1}{S}\int_0^S \varphi(\tau)\,d\tau \qquad (3.9)$$

if $T > \frac{2S\|f\|^2}{\epsilon^2}$. Now invoke Lemma 3.13 to get $\lambda_1,\ldots,\lambda_n$, and for $\rho > 0$, an S such that $\frac{1}{S}\int_0^S |e^{i\lambda\tau}-1|^2\varphi(\tau)\,d\tau > \frac{1}{18S}\int_0^S \varphi(\tau)\,d\tau$ if $|\lambda-\lambda_i| \ge \rho$. Then fix this S to get

$$\frac{T}{2\pi}\int_{-\infty}^{\infty} |a_T(\lambda)|^2 \frac{1}{S}\int_0^S |e^{i\lambda\tau}-1|^2\varphi(\tau)\,d\tau\,d\lambda \ge \frac{T}{2\pi}\int_U |a_T(\lambda)|^2 \frac{1}{S}\int_0^S |e^{i\lambda\tau}-1|^2\varphi(\tau)\,d\tau\,d\lambda$$

$\ge \frac{T}{2\pi}\int_U |a_T(\lambda)|^2 \frac{1}{18S}\int_0^S \varphi(\tau)\,d\tau\,d\lambda$. Now we combine this with 3.9 to get

$$\frac{T}{2\pi}\int_U |a_T(\lambda)|^2 d\lambda \le 36\epsilon^2, \text{ if } T \ge \frac{2S\|f\|^2}{\epsilon^2}.$$

Theorem 3.15. (Parseval's Equation) For any $f \in AP(E^n)$, $M(|f|^2) = \sum |a(\lambda)|^2$.

Proof: Note that we have reduced this theorem to the inequality 3.7. If $\epsilon > 0$ is given we invoke Lemma 3.14 to get $\lambda_1,\ldots,\lambda_N$. We invoke Lemma 3.11 N times to get a T_0 and $\delta > 0$ so that

$$\frac{T}{2\pi}\int_{\lambda_i-\delta}^{\lambda_i+\delta} |a_T(\lambda)|^2 d\lambda < |a(\lambda_i)|^2 + \epsilon/N$$

when $T > T_0$. For this δ we can find a T_1 so that $\frac{T}{2\pi}\int_U |a_T(\lambda)|^2 d\lambda < \epsilon$ when $T > T_1$ and $U \equiv \{\lambda \mid |\lambda-\lambda_i| \geq \delta\}$. If $T \geq \max(T_1, T_0)$ we have

$$\frac{T}{2\pi}\int_{-\infty}^{\infty} |a_T(\lambda)|^2 d\lambda \leq \frac{T}{2\pi}\int_U |a_T(\lambda)|^2 d\lambda + \sum_{i=1}^{N} \frac{T}{2\pi}\int_{\lambda_i-\delta}^{\lambda_i+\delta} |a_T(\lambda)|^2 d\lambda$$

$$\leq \sum_{i=1}^{N} |a(\lambda_i)|^2 + 2\epsilon \leq \sum |a(\lambda)|^2 + 2\epsilon .$$

This completes the proof of the theorem.

6. Approximation Theorem. We base the proof of the Approximation Theorem on the proof for the same theorem for periodic functions.

The proof for periodic functions is to show that the Cesaro-means of the partial sums converge uniformly to the (continuous) periodic function. To formulate the central idea, let $f(x) \sim \sum_{\nu=-\infty}^{\infty} a_\nu e^{i\nu T x}$, then the Cesaro means are given by the formula

$\sigma_n(x) = \sum_{|\nu|<n} \left(1 - \frac{|\nu|}{n}\right) a_n e^{i\nu T x}$. Now $a_n = M_t(f(t)e^{-i\nu T t})$ so that

$a_n e^{i\nu T x} = M_t(f(t)e^{-i\nu T(x-t)}) = M_t(f(t+x)e^{-i\nu T t})$ so that

$\sigma_n(x) = M_t(f(t+x)K_n(tT))$ where

$$K_n(t) = \sum_{|\nu|<n} \left(1 - \frac{|\nu|}{n}\right) e^{-i\nu t} = \frac{1}{n}\left(\frac{\sin\frac{n}{2}t}{\sin\frac{t}{2}}\right)^2. \qquad (3.10)$$

The main properties of $K_n(t)$ are

i) $K_n(t) \geq 0$ and

ii) $M(K_n) = 1$.

Property (i) is clear from the second representation and (ii) from the first since $M(K_n)$ is the constant term. Roughly, the proof then is as follows,

$$|f(x) - \sigma_n(x)| = M_t(f(x)K_n(Tt) - f(x+t)K_n(Tt))$$

$$\leq M_t(|f(x) - f(x+t)|K_n(Tt))$$

and the idea is that for $|t|$ small the first factor is small and for $|t|$ bounded away from 0, $K_n(Tt) \to 0$ as $n \to \infty$.

To do the almost periodic case, one takes products of K_n's, one for every periodic component of the almost periodic function. We want to base our proof on u.a.p. families. So first look at the above periodic case. It is easy to see that the $\sigma_n(x)$ are a u.a.p. family. Indeed, $|\sigma_n(x)| \leq \|f\| M_t(K_n(tT)) = \|f\|$ and $|\sigma_n(x+\tau) - \sigma_n(x)| \leq M_t(|f(t+x+\tau) - f(t+x)|K_n(Tt))$ $\leq v_f(\tau) M_t(K_n(Tt)) = v_f(\tau)$ so that $v_{\sigma_n}(\tau) \leq v_f(\tau)$. Hence some subsequence of the σ_n's converges uniformly to an almost periodic function g. But clearly, the formal fourier series of σ_n converge to the formal fourier series of f. So by Theorem 3.7 and the uniqueness of fourier series of periodic functions $f = g$ and the proof is complete.

To set the problem, let $f(x,t)$, $x \in K$, be a u.a.p. family, with exponents Λ. Enumerate Λ by $\lambda_1, \lambda_2, \ldots$ in any way. There are a set of real numbers $\{\beta_1, \beta_2, \ldots\}$ which are linearly independent over the rationals and such that each $\lambda_i = \nu_1\beta_1 + \cdots + \nu_n\beta_n$ where the ν_i are rational numbers. The β's are called a base for Λ. Note that in general they are not in Λ.

The second introductory notion that we need is that of the Bochner-Fejer polynomials. Let $\alpha_1, \ldots, \alpha_n$ be real numbers which are linearly independent over the rationals. Let $m_1, \ldots, m_n$ be positive integers and consider the trigonometric polynomials

$$K(m,\alpha,t) = K_{m_1}(\alpha,t)\, K_{m_2}(\alpha,t) \cdots K_{m_n}(\alpha_n t)$$

$$= \sum_{|\nu_i| < m_i} \left(1 - \frac{|\nu_1|}{m_1}\right) \cdots \left(1 - \frac{|\nu_n|}{m_n}\right) e^{-i(\nu_1\alpha_1 + \cdots + \nu_n\alpha_n)t}$$

where $K_{m_i}(\alpha_i t)$ are given by 3.10, $m = (m_1, \ldots, m_n)$, ν_i integral, and $\alpha = (\alpha_1, \ldots, \alpha_n)$. It follows that $K(m,\alpha,t) \geq 0$ and $M(K(m,\alpha,t)) = 1$. The latter follows since $\nu_1\alpha_1 + \cdots + \nu_n\alpha_n = 0$ only if $\nu_i = 0$ $i = 1, \ldots, n$ by the linear independence of the α's. We can then form the Bochner-Fejer polynomials corresponding to f

$$\sigma(m,\alpha,s) = M_t(f(s+t)K(m,\alpha,t))$$

$$= \sum_{|\nu_i| < m_i} \left(1 - \frac{|\nu_1|}{m_1}\right) \cdots \left(1 - \frac{|\nu_n|}{m_n}\right) a(f, \nu_1\alpha_1 + \cdots + \nu_n\alpha_n) e^{i(\nu_1\alpha_1 + \cdots + \nu_n\alpha_n)s} \tag{3.11}$$

It follows that the exponents of $\sigma(m,\alpha,x)$ are contained in the exponents of f. These polynomials may be written in the form

$\sigma(m,\alpha,x) = \sum d(m,\alpha,n)\ a(f,\lambda_n)e^{i\lambda_n x}$ where the numbers $d(m,\alpha,n)$ satisfy the following properties, (i) $0 \leq d(m,\alpha,n) \leq 1$, (ii) for fixed (m,α) only a finite number are non-zero, and (iii) the numbers $d(m,\alpha,n)$ depend only on (m,α,n) and the λ_n but not on f or $a(f,\lambda_n)$.

Lemma 3.16. Let $f(x,t)$ for $x \in K$ be a u.a.p. family. Then f together with the Bochner-Fejer polynomials are a u.a.p. family.

Proof: Let v be the translation function for $f(x,t)$, $x \in K$, and (m,α) be arbitrary. Then according to (3.11) $\sigma(m,\alpha,x,s+\tau) - \sigma(m,\alpha,x,s) = M_t((f(x,s+\tau+t) - f(x,s+t))K(m,\alpha,t))$ so that $v_\sigma(\tau) \leq v(\tau)M_t(K(m,\alpha,t)) = v(\tau)$ since $K(m,\alpha,t) \geq 0$ and $M_t(K(m,\alpha,t)) = 1$. Here $\sigma(m,\alpha,x,s) = M_t(f(x,s+t)K(m,\alpha,t))$.

Theorem 3.17. (Approximation Theorem) Let $f(x,t)$, $x \in K$ be a u.a.p. family, then for every $\epsilon > 0$, there is a trigonometric polynomial $p(x,t)$ such that $\|f(x,t) - p(x,t)\| < \epsilon$ for $x \in K$. Furthermore, p may be selected to have exponents contained in the exponents of f.

Proof: In view of Lemma 3.16 and Theorem 3.10 it is sufficient to exhibit a sequence of Bochner-Fejer polynomials that converge in the mean to f. As above we enumerate the exponents of f and find a base $\{\beta_1,\beta_2,\ldots\}$. By Parseval's equation, $\sum_{n=1}^{\infty} |a(f,\lambda_n,x)|^2$ converges uniformly on K. Let N be picked so that

$$\sum_{n=N+1}^{\infty} |a(f,\lambda_n,x)|^2 < \epsilon/2. \tag{3.12}$$

Consider $\lambda_1,\ldots,\lambda_N$ and let p be chosen so that $\lambda_i = r_1^{(i)}\beta_1 + \ldots + r_p^{(i)}\beta_p$ for $i = 1,\ldots,N$, where certain of $r_j^{(i)}$ may be zero, but all are rational numbers. Let q be a common multiple of the denominators of all the numbers $r_j^{(i)}$. Thus

$$\lambda_i = s_1^{(i)} \frac{\beta_1}{q} + \ldots + s_p^{(i)} \frac{\beta_p}{q}$$

where the $s_j^{(i)}$ are now integers. Let S be the $\max_{i,j} |s_j^{(i)}|$. Find a positive number ρ so that

$$\rho^2 \sum_{n=1}^{N} |a(f,\lambda_n,x)|^2 < \epsilon/2 \tag{3.13}$$

and then a number M so that

$$(1 - \frac{S}{M})^p > 1-\rho. \tag{3.14}$$

Now define $\alpha_i = \frac{\beta_i}{q}$ and $m_i > M$ $i = 1,\ldots,p$ and consider $\sigma(m,\alpha,x,t)$, the Bochner-Fejer polynomial of $f(x,t)$. Specifically,

$$\sigma(m,\alpha,x,t) = \sum_{n=1}^{N} \left(1 - \frac{|s_1^{(n)}|}{m_1}\right) \cdots \left(1 - \frac{|s_p^{(n)}|}{m_p}\right) a(f,\lambda_n,x) e^{i\lambda_n t}$$

$$+ \sum_{n=N+1}^{\infty} d(m,\alpha,n) a(f,\lambda_n,x) e^{i\lambda_n t} .$$

Then by Parseval's equation, (3.12), (3.13), (3.14), and $0 \leq d \leq 1$,

$$M_t(|f(x,t) - \sigma(m,\alpha,x,t)|^2) = \sum_{n=1}^{N} \left\{1 - \left[\left(1 - \frac{|s_1^{(n)}|}{m_1}\right) \cdots \left(1 - \frac{|s_p^{(n)}|}{m_p}\right)\right]\right\}^2$$

$$\times |a(f,\lambda_n,x)|^2 + \sum_{n=N+1}^{\infty} (1 - d(m,\alpha,n))^2 \; |a(f,\lambda_n,x)|^2$$

$$\leq \rho^2 \sum_{n=1}^{N} |a(f,\lambda_n,x)|^2 + \sum_{n=N+1}^{\infty} |a(f,\lambda_n,x)|^2 < \epsilon \; .$$

Corollary 3.18. The Almost Periodic Functions are precisely those that can be uniformly approximated by trigonometric polynomials.

Proof: By Theorem 3.17, a.p. functions can be so approximated. On the other hand, $ae^{i\lambda t} \in AP$ for all a and λ. Furthermore, AP is closed under finite sums and uniform approximation by Theorem 1.9, so all functions that can be uniformly approximated by trigonometric polynomials are in AP.

It is this result, which parallels the periodic case, that shows that the almost periodic functions are an appropriate class to study.

7. Differentiation and Integration of Fourier Series. It is of interest to consider the question of formal differentiation and integration of fourier series of almost periodic functions. Suppose in fact f and f' are almost periodic. Then one can compute the fourier series of f' in the usual way. Indeed,

$$a(f',\lambda) = \lim_{T\to\infty} \frac{1}{T}\int_0^T f'(t)e^{-i\lambda t}\, dt$$

$$= \lim_{T\to\infty} \left[\frac{e^{-i\lambda T}f(T) - f(0)}{T} + i\lambda \frac{1}{T}\int_0^T f(t)e^{-i\lambda t}dt\right]$$

$$= i\lambda\ a(f,\lambda)$$

since f is bounded. In particular $a(f',0) = 0$ and so the fourier series of f' is the formal derivative of the fourier series of f.

On the other hand if f and $\int_0^t f(s)ds$ are both almost periodic, then by the above $a(f,0) = 0$ is necessary and it follows that the fourier series of $\int_0^t f(s)ds$ is $a_0 + \sum \frac{a(f,\lambda)}{i\lambda} e^{i\lambda t}$. An immediate question is the sufficiency of $a(f,0) = 0$ for $\int_0^t f$ to be almost periodic. In fact it is not. To see this, consider the series $f(t) = \sum_{n=1}^{\infty} \frac{1}{n^2} e^{it/n^2}$. The series converges uniformly so its sum, f is almost periodic. However, if $\int_0^t f$ were almost periodic, then

$$\int_0^t f(s)ds \sim a_0 + \sum_{n=1}^{\infty} e^{it/n^2} .$$

This is not possible since the coefficients are not square summable, violating Parseval's equation. We shall see later that if $|\lambda_n| \geq \alpha > 0$, then $\int_0^t f$ is almost periodic. This is the only known simple condition on the fourier series which yields the almost periodicity of $\int_0^t f$, except for the obvious condition that

$$\sum \left|\frac{a(f,\lambda)}{\lambda}\right| < \infty .$$

8. Differentiability of $a(f,\lambda,x)$. We have seen that $a(f,\lambda,x)$ has the continuity property of $f(x,t)$. We would like to extend this to show that differentiability properties of f carry over to $a(f,\lambda,x)$.

The kind of situation we are interested in is the following. By $\int_{x_0}^{x} f(\xi,t)d\xi$ we will mean coordinate wise integration and by $\frac{\partial}{\partial x_i} f(x,t)$ we will mean taking the indicated partial derivative of each component of the vector $f(x,t)$. So the kind of question we will ask can be reduced to the same question for scalar valued functions defined on $E^n \times R$. A typical question might be, "if $\frac{\partial f}{\partial x_i}(x,t)$ is in $AP(E)$, does $\frac{\partial}{\partial x_i} a(f,\lambda,x)$ exist?". The answer is quite simple.

Theorem 3.19. Let $f(x,t)$ and $\frac{\partial f}{\partial x_i}(x,t)$ be almost periodic uniformly for x in compact sets, then $\frac{\partial}{\partial x_i} a(f,\lambda,x)$ exists and

$\frac{\partial f}{\partial x_i} a(f,\lambda,x) = a(\frac{\partial f}{\partial x_i},\lambda,x)$.

Proof: Let K be a product $[a,b] \times \cdots \times [a,b]$. Then $a(\frac{\partial f}{\partial x_i},\lambda,x) = \lim_{T\to\infty} \frac{1}{T}\int_0^T \frac{\partial f}{\partial x_i}(x,t)e^{-i\lambda t}dt$ uniformly for $x \in K$. If $x_0 = (y_1,\ldots,y_i,\ldots,y_n)$ and $x = (y_1,\ldots,z_i,\ldots,y_n)$ are in K then

$$\int_{x_0}^{x} a(\frac{\partial f}{\partial x_i},\lambda,(y_1,\ldots,\xi_i,\ldots,y_n))d\xi_i =$$

$$\lim_{T\to\infty} \frac{1}{T}\int_0^T \left(\int_{x_0}^{x} \frac{\partial f}{\partial x_i}\big((y_1,\ldots,\xi_i,\ldots,y_n),t\big)d\xi_i\right)e^{-i\lambda t}dt$$

$$= \lim_{T\to\infty} \frac{1}{T}\int_0^T [f(x,t) - f(x_0,t)]e^{-i\lambda t}dt$$

$$= a(f,\lambda,x) - a(f,\lambda,x_0)$$

where the change of order of integration is valid by the uniform existence of $a(\frac{\partial f}{\partial x_i},\lambda,x)$. Thus we have

$$a(f,\lambda,x) = a(f,\lambda,x_0) + \int_{x_0}^{x} a(\frac{\partial f}{\partial x_i},\lambda,(y_1,\ldots,\xi_i,\ldots,y_n))d\xi_i \qquad (3.15)$$

so that by the continuity of the integrand, $\frac{\partial}{\partial x_i} a(f,\lambda,x)$ exists and is equal to the integrand.

Questions about the integration of fourier coefficients can be handled by the identity (3.15) and its obvious generalizations to gradients and higher derivatives. In particular, if $E^n = C$ then the results of Theorem 3.19 show that the Cauchy-Riemann equations on a are the same as on f so that if $f(z,t)$ is analytic in z then so is $a(f,\lambda,z)$.

<u>Corollary 3.20</u>. Suppose $f(z,t)$ is in $AP(C)$ uniformly for z in a domain Ω. If $f(z,t)$ is analytic in z in Ω, then $a(f,\lambda,z)$ is analytic for $z \in \Omega$.

<u>Proof</u>: We observe that if $f(z,t)$ is analytic and u.a.p. in Ω then on compact subsets, f is uniformly bounded. By writing down the Cauchy integral formula for f in terms of a fixed Jordan curve, it becomes apparent that the derivative of f with respect to z is uniformly continuous in t. From $f'(z,t) = \frac{1}{2\pi i}\int_C \frac{f(\zeta,t)d\zeta}{(\zeta-z)^2}$

one invokes the uniform continuity of $f(\zeta,t)$ on $C \times R$ to get the uniform continuity of $f'(z,t)$ in t by the simple estimate

$$|f'(z,t+\tau) - f'(z,t)| \leq \frac{1}{2\pi} \sup_{(\zeta,t)} \frac{|f(\zeta,t+\tau) - f(\zeta,t)|}{\min_{\zeta} |\zeta - z|^2} (\text{length } (C)). \quad (3.16)$$

Thus $f'(z,t)$ is almost periodic in t, and in fact uniformly for z is small compact sets, as (3.16) shows. By Theorem 3.19
$\frac{d}{dz} a(f,\lambda,z) = a(f',\lambda,z)$ so $a(f,\lambda,z)$ is analytic.

9. Notes. The harmonic analysis of almost periodic functions should be motivated by the approximation theorem.

Theorem 3.1 is essentially Bohr's original proof. Note that the proof of Bessel's inequality parallels that for ordinary Fourier series or for abstract Hilbert space. Indeed, using the mean value for a norm, $AP(R)$ is a pre-Hilbert space, not being complete. Its completion is an example of a Hilbert space with the dimension being the cardinality of R.

The point of Corollary 3.4 is that even though there are uncountably many almost periodic functions, each with a countable set of exponents, the aggregate is still countable. This is the strength of the u.a.p. hypothesis.

The results on mean convergence and the proof of the Approximation Theorem are again due to Bochner and the methods closely parallel those for the periodic case.

Our exposition of the Parseval relation is that of Jessen, On the proofs of the Fundamental Theorem on Almost Periodic Functions, Det. KGL Danske Viden. Selskob Mat-Fys. Meddel. XXV(1949), 1-12.

The comment that there are no known simple conditions on the fourier series to get the almost periodicity of the integral shows that the problem deserves more study.*

It is possible to reverse the order of the proofs in this chapter somewhat. One can prove the Approximation Theorem first and then Parseval's equation is an easy consequence. Such an exposition can be found in the book of Corduneau**, Almost Periodic Functions, Interscience, 1968.

One should mention that with the exception of the results on mean convergence, all of the results in this chapter were first proved by Bohr.

Most of the results of this chapter hold for a.p. functions on groups or semi-groups into Banach spaces, see comment on 2.13.

* See Meisters [547].
** (544)

Chapter 4

Modules and Exponents

1. Introduction. We center our attention on the question of the connection between the translates of an a.p. function, the ϵ-translation numbers and the exponents. In its simplest form, suppose we know f is periodic and that the existence of $T_\alpha f = f$ uniformly implies the uniform existence of $T_\alpha g = g$. Certainly, we can conclude that g is almost periodic. Can more be said? Take α_n = period of f, then $T_\alpha f = f$ and $T_\alpha g = g$ so that g is periodic and its period divides that of f. Although it is easy to see that f and g need not have any common exponents, it must be true that the additive group of possible exponents of g is a subset of the additive group of possible exponents of f, since (period f) = n (period g).

A second topic that we want to discuss is the summation of fourier series by convolution with fourier transforms of a type that are somewhat different from the Bochner - Fejer polynomials.

2. Translation sets. We first explore the relationship between the ϵ-translation set of a function and the partial sums of its fourier series.

Theorem 4.1. If $f \in AP(E)$ and

$$f(x) \sim \sum_{n=1}^{\infty} a_n e^{i\lambda_n x}$$

then for every positive integer N and positive number $\delta < \pi$, there is an $\epsilon > 0$ so that

$$T(f,\epsilon) \subset \{\tau \mid |\lambda_n \tau| < \delta (\mathrm{mod}\ 2\pi), n = 1,\ldots,N\}.$$

Proof. For $\delta < \pi$ the inequality $|e^{i\lambda_n \tau} - 1| < |e^{i\delta} - 1| = \rho$ is equivalent to $|\lambda_n \tau| < \delta (\mathrm{mod}\ 2\pi)$. Let $A = \min\{|a_1|,\ldots,|a_N|\}$ and $\epsilon = \frac{A}{2}\rho$. If $\tau \in T(f,\epsilon)$ then $|a_n(e^{-i\lambda_n \tau} - 1| =$

$|M_t\{(f(t+\tau) - f(t))e^{-i\lambda_n t}\}| \leq M_t\{|f(t+\tau) - f(t)|e^{-i\lambda_n t}\} \leq \epsilon < A\rho.$

Therefore $|e^{-i\lambda_n \tau} - 1| < \rho$.

Note that the set $\{\tau \mid |\lambda_n \tau| < \delta (\mathrm{mod}\ 2\pi)\}$ is $T(e^{i\lambda_n t}, \rho)$ so that $T(f,\epsilon)$ is actually contained in a finite intersection of periodic sets. The next theorem is a sort of converse.

Theorem 4.2. Suppose $f \in AP(E)$ and

$$f(x) \sim \sum_{n=1}^{\infty} a_n e^{i\lambda_n x}.$$

For every $\epsilon > 0$, there is a positive integer N and $0 < \delta < \pi$ such that

$$\{\tau \mid |\lambda_n \tau| < \delta (\mathrm{mod}\ 2\pi) n = 1,2,\ldots,N\} \subset T(f,\epsilon).$$

Proof. We use the Approximation Theorem. Select $p(t) = \sum_{n=1}^{N} b_n e^{i\lambda_n x}$ with exponents of p contained in those of f such that $\|f - p\| < \epsilon/3$. Let $A = \sum_{n=1}^{N} |b_n|$, select $\delta_1 < \epsilon/3A$, and define $\delta < \pi$ by $|e^{i\delta} - 1| = \delta_1$. If $|e^{i\lambda_n\tau} - 1| < \delta_1, n = 1,\ldots,N$ then $|\lambda_n\tau| < \delta(\mathrm{mod}\ 2\pi), n = 1,\ldots,N$. Therefore $|p(t+\tau) - p(t)| \leq \sum_{n=1}^{N} |b_n||e^{i\lambda_n\tau} - 1| < \delta_1 A < \epsilon/3$. Now by an easy triangle inequality argument, $T(f,\epsilon/3) \subset T(t,\epsilon)$.

3. Kronecker's Theorem. We wish to give a proof of a classical theorem of Kronecker. This theorem is a number theoretic theorem but one can use almost periodic functions to prove it. Originally, it was used to prove some facts about almost periodic functions. Such interrelationships are quite satisfying. We will consider the system of inequalities

$$|\lambda_k\tau - \theta_k| < \delta(\mathrm{mod}\ 2\pi), k = 1,\ldots,n \qquad (4.1)$$

where $\lambda_k, \theta_k, \delta$ are real numbers.

Theorem 4.3. (Kronecker) The system 4.1 has a solution τ for any $\delta > 0$ if and only if for all integral m_k

$$\sum_{k=1}^{n} m_k\lambda_k = 0 \quad \text{implies} \quad \sum_{k=1}^{n} \theta_k\lambda_k \equiv 0(\mathrm{mod}\ 2\pi).$$

Proof: If there is a solution τ for every $\delta > 0$, there are integers n_k so that

$$\delta_k = \lambda_k\tau - \theta_k - 2\pi n_k$$

satisfies $|\delta_k| < \delta$ for $k = 1,\ldots,n$. If m_k^j are integers such that $\sum_{k=1}^{n} m_k\lambda_k = 0$ then $\sum_{k=1}^{n} m_k\delta_k = \sum_{k=1}^{n} - m_k\theta_k - \sum_{k=1}^{n} 2\pi n_k m_k$ and so $|\sum_{k=1}^{n} m_k\theta_k + 2\pi \sum_{k=1}^{n} m_k n_k| < \delta \sum_{k=1}^{n} |m_k|$. Since δ is arbitrary $\sum_{k=1}^{n} m_k\theta_k = - 2\pi(\sum_{k=1}^{n} m_k n_k)$.

- Conversely, suppose λ_k, θ_k are given and consider the function in AP(C) defined by

$$f(t) = 1 + \sum_{k=1}^{n} e^{i(\lambda_k t - \theta_k)} = 1 + \sum_{k=1}^{n} e^{-i\theta_k} e^{i\lambda_k t}.$$

If m is a positive integer we want to consider $(f)^m$. If $|\lambda_k t - \theta_k|$ is close to zero then f should be close to $(n+1)$. If this fails then it fails even more so for $(f)^m$ and so should be easier to detect. Now $(f)^m$ is a trigonometric polynomial of the form

$$f^m(x) = \sum a_j e^{i\beta_j x} \tag{4.2}$$

where $\beta_j = \sum_{\text{finite}} m_k^j\lambda_k$, $\sum m_k^j = m$, m_k^j are non-negative integers, and j runs over the ordered partitions of m. Note that if $\beta_j = \beta_{j'}$, then $\sum (m_k^j - m_k^{j'})\lambda_k = 0$ and so $\sum (m_k^j - m_k^{j'})\theta_k = 0 \bmod 2\pi$. The coefficients a_j and $a_{j'}$ are integral multiples of complex numbers with the same argument, namely $e^{-i(\Sigma m_k^j\theta_k)}$. Thus in 4.2

we collect all terms that have $e^{i\beta_j x}$ as a factor and write

$$f^m(x) = \sum b_j e^{i\beta_j x} \tag{4.3}$$

with distinct β_j. Now $\sum |a_i|$, with the sum take over those i's for which $\beta_i = \beta_j$, is equal to $|\beta_j|$ since all a_i have the same argument. But consider where the a_j came from. To find $f^m(x)$ one looks at $(1 + u_1 + \ldots + u_n)^m = \sum b(m_1,\ldots,m_n) u_1^{m_1},\ldots,u_n^{m_n}$ and substitutes $u_j = e^{i(\lambda_j t - \theta_j)}$. Thus

$|a_j| = |b(m_1,\ldots,m_n)| = b(m_1,\ldots,m_n)$ and

$$\sum_{\text{distinct } \beta_j} |b_j| = \sum_j |a_j| = \sum b(m_1,\ldots,m_n) = (1 + 1 + \ldots + 1)^m = (n+1)^m.$$

Now if $\|f\| < n+1$ say $\|f\| = \theta(n+1)$ $0 < \theta < 1$ then $|b_j| \le \|f^{(m)}\| \le \theta^m(n+1)^m$. Thus $(n+1)^m = \sum |b_j| \le \theta^m(n+1)^m$ (number of distinct terms in 4.3) $\le \theta^m(n+1)^m(m+1)^n$. But the inequality $1 \le \theta^m(m+1)^n$ is not possible for large m. This proves that $\|f\| = n+1$. Hence for every $\epsilon > 0$, there is a τ such that $|e^{i(\lambda_k\tau - \theta_k)}| > 1 - \epsilon$, $k = 1,\ldots,n$, as desired.

4. Module containment. We are now in the position to make a definitive statement on the relationships between the exponents of two almost periodic functions.

Definition 4.4. Let $f \in AP(E)$, then the module of f, Mod(f) is the smallest additive group of real numbers that contains the exponents of f.

Theorem 4.5. The following statements are equivalent for f and $g \in AP(E)$

(i) $\text{Mod}(f) \supset \text{Mod}(g)$;

(ii) for every $\epsilon > 0$, there is a $\delta > 0$ so that $T(f,\delta) \subset T(g,\epsilon)$;

(iii) $T_\alpha f$ exists implies $T_\alpha g$ exists (any sense);

(iv) $T_\alpha f = f$ implies $T_\alpha g = g$ (any sense);

(v) $T_\alpha f = f$ implies there is $\alpha' \subset \alpha$ so that $T_{\alpha'} g = g$ (any sense).

Comment: Any sense means that the convergence may be interpreted as uniform, uniform on compact sets, pointwise, or in the mean. They all are equivalent by Theorems 2.4 and 3.10.

Proof: The plan of the proof is to show that (i) $\Rightarrow$ (ii) $\Rightarrow$ (iv) $\Rightarrow$ (v) $\Rightarrow$ (ii) $\Rightarrow$ (i) and (iii) $\Leftrightarrow$ (iv). (i) $\Rightarrow$ (ii). If $\epsilon > 0$ is given and $\text{Mod}(g) \subset \text{Mod}(f)$, let $f(t) \sim \sum a_k e^{i\lambda_k t}$ and $g(t) \sim \sum b_k e^{im_k t}$. By Theorem 4.2 there is an N and $\delta < \pi$ so that $\{\tau \mid |m_k\tau| < \delta (\text{mod } 2\pi),\ k = 1,\dots,N\} \subset T(g,\epsilon)$. Now $m_k = \sum_{i=1}^{K} \alpha_i^{(k)} \lambda_i, k = 1,\dots,N$ and the $\alpha_i^{(k)}$ are integers. Let $A = \max|\alpha_i^{(k)}|$. If $|\lambda_n\tau| < \frac{\delta}{KA} \text{mod}(2\pi)$ $n = 1,\dots,K$, then $-\frac{\delta}{KA} < \lambda_n\tau - 2\pi j_n < \frac{\delta}{KA}, n = 1,\dots,K$ for certain integers j_n. Consequently

$$- \delta/K < \lambda_n\tau\alpha_n^{(k)} - 2\pi\alpha_n^{(k)} j_n < \delta/K,\ n = 1,\dots,K.$$

Adding one gets $-\delta < \tau m_k - 2\pi M < \delta$ for M an integer, and $k = 1,\dots,N$. Hence

$S = \{\tau | \; |\lambda_n \tau| < \frac{\delta}{KA} (\text{mod } 2\pi), \; n = 1,\dots,K\} \subset T(g,\epsilon)$. By Theorem 4.1, there is a $\rho > 0$ so that $T(f,\rho) \subset S$.

(ii) $\Rightarrow$ (iv) Let $T_\alpha f = f$ uniformly and suppose $\epsilon > 0$. There is a $\delta > 0$ so that $T(f,\delta) \subset T(g,\epsilon)$. Now $|f(t+\alpha_n) - f(t)| < \delta$ for large n so that $\alpha_n \in T(g,\epsilon)$ for large n. That is $|g(t+\alpha_n) - g(t)| < \epsilon$ for $n \geq N$. This implies $T_\alpha g = g$ uniformly.

(iv) $\Rightarrow$ (v) proof is obvious.

(v) $\Rightarrow$ (ii) If (ii) is not true there is an $\epsilon > 0$ so that $T(f,1/n) \not\subset T(g,\epsilon)$ for all n. Select $\alpha_n \in T(f,1/n)$ and $\alpha_n \notin T(g,\epsilon)$. Then $|f(t+\alpha_n) - f(t)| < 1/n$ implies that $T_\alpha f = f$, but $\alpha_n \notin T(g,\epsilon)$ implies that $\|g(t+\alpha_n) - g(t)\| \geq \epsilon$ for all n.

(ii) $\Rightarrow$ (i) This is by far the deepest implication and is an application of Kronecker's Theorem. We suppose there is a number $M \in \text{Mod}(g)\backslash\text{Mod}(f)$. Two cases arise which we can handle simultaneously with parallel proofs. Case 1 is when $kM \in \text{Mod}(f)$ for some non-zero integer k. Case 2 is when this is not the situation. We first select an integer $k_o \geq 2$. In Case 1, let k_o be the absolute value of the smallest $k \neq 0$ for which $kM \in \text{Mod}(f)$. Note $k_o \neq 1$. In Case 2 take $k_o = 2$. Now fix $0 < \delta_1 < \pi/k_o$. By Theorem 4.1, there is an $\epsilon > 0$ so that $T(g,\epsilon) \subset \{\tau | \; |M\tau| < \delta_1 (\text{mod } 2\pi)\}$ and by (ii) there is a δ_2 so that $T(f,\delta_2) \subset T(g,\epsilon)$. Now apply Theorem 4.2 to get N, δ_3 so that

$$\{\tau \mid |\lambda_j \tau| < \delta_3 (\text{mod } 2\pi),\ j = 1,\ldots,N\} \subset T(f,\delta_2), \quad (\lambda_j \in \exp(f))$$

We may assume that $\delta_3 < \delta_1$ and we have

$$\{\tau \mid |\lambda_j \tau| < \delta_3 (\text{mod } 2\pi)\ j = 1,\ldots,N\} \subset \{\tau \mid |M\tau| < \delta_1 (\text{mod } 2\pi)\}. \qquad (4.4)$$

We want to apply Kronecker's Theorem to the set λ_j and $\theta_j = 0$ $j = 1,\ldots,N$, $M = \lambda_o$ and θ_o, where $\theta_o = \frac{2\pi}{k_o}$ for Case 1 and $\theta_o = \pi$ for Case 2. In order to do this, we investigate when $\ell_o M + \sum_{i=1}^{N} \ell_i \lambda_i = 0$ for integral ℓ_i. The claim is that

$\ell_o \theta_o = 0 (\text{mod } 2\pi)$. This is easy to see in Case 2, since $\ell_o = 0$. In Case 1, we claim that $\ell_o = \rho k_o$ for ρ an integer. Indeed, if $\rho_o = qk_o + r$ with q and r integral and $0 \leq r < k_o$, then $qk_o M + rM + \sum_{i=1}^{N} \ell_i \lambda_i = 0$. But $k_o M = \sum_{i=1}^{K} t_i \lambda_i$ t_i, integral implies that $rM \in \text{Mod}(f)$. This is only possible if $r = 0$ since k_o is minimal. Thus $\ell_o = \rho k_o$ so $\ell_o \theta_o = \rho k_o \frac{2\pi}{k_o} = 2\pi\rho$. This shows that λ_k, θ_k satisfy half of Kronecker's Theorem. Therefore there is a τ such that $|\lambda_j \tau| < \delta_3 (\text{mod } 2\pi)$, $j = 1,\ldots,N$ and $|M\tau - \theta_o| < \delta_3 (\text{mod } 2\pi)$. By 4.4 $|M\tau| < \delta_1 (\text{mod } 2\pi)$. The last two inequalities show that

$$-\delta_1 < 2m\pi - M\tau < \delta_1$$

and
$$-\delta_3 < M\tau - \theta_o + 2n\pi < \delta_3$$

for certain integral n and m. Adding we get

$$-(\delta_1+\delta_3) < 2(m+n)\pi - \theta_o < \delta_1 + \delta_3 .$$

In Case 1 $\theta_o = \frac{2\pi}{k_o} < \pi$, and using $\delta_3 < \delta_1 < \pi/k_o$ one gets $0 < 2(m+n)\pi < \frac{4\pi}{k_o} \leq 2\pi$. In Case 2 $\theta_o = \pi$ and we use $\delta_3 < \delta_1 < \pi/2$ to get $-\pi < [2(m+n)-1]\pi < \pi$. In either case we have a contradiction.

(iv) ⇒ (iii) Suppose $T_\alpha f$ exists uniformly and suppose $T_\alpha g$ does not. Let $\alpha_1 \subset \alpha$ and $\alpha_2 \subset \alpha$ so that $\|g(t+\alpha_{1n}) - g(t+\alpha_{2n})\| \geq \epsilon_o > 0$ then $\|g(t+\alpha_{1n}-\alpha_{2n}) - g(t)\| \geq \epsilon_o$ for all n. But $\|f(t+\alpha_{1n}) - f(t+\alpha_{2n})\| \to 0$ as $n \to \infty$ so $\|f(t+\alpha_{1n}-\alpha_{2n}) - f(t)\| \to 0$ as $n \to \infty$. That is, $T_{\alpha_1-\alpha_2} f = f$. By (4) $T_{\alpha_1-\alpha_2} g = g$ which is a contradiction.

(iii) ⇒ (iv) Let $T_\alpha f = f$ and $g^* = T_\alpha g$. This exists by (iii). Consider $\beta = (\alpha_1, 0, \alpha_2, 0, \alpha_3, \ldots)$. Then $T_\beta f = f$ so $T_\beta g = g_1$ by (iii). Now $\alpha \subset \beta$ so $g^* = T_\alpha g = T_\beta g = g_1$. Also the zero sequence $0 \subset \beta$ so $g = T_o g = T_\beta g = g_1$. Therefore $g = g^*$ as required.

<u>Theorem 4.6</u>. If $f(x,t)$ is almost periodic uniformly for $x \in K$, and φ is almost periodic such that $\varphi(t) \in K$ for all t, then $\mathrm{mod}(\varphi) \subset \mathrm{mod}(f)$ implies that $\mathrm{mod}\, f(\varphi(t),t) \subset \mathrm{mod}(f)$.

<u>Proof</u>: Let $T_\alpha f(x,t) = f(x,t)$ uniformly on $K \times R$. Then $T_\alpha \varphi = \varphi$ uniformly by Theorem 4.5 (iv). Then it is easy to see

that $T_\alpha f(\varphi(t),t) = f(\varphi(t),t)$ and again by Theorem 4.5 (i) $\mathrm{mod}(f(\varphi,t)) \subset \mathrm{mod}(f)$.

5. Convolution by Fourier Transforms. One way to generate new almost periodic functions from a given one is to convolute it with other functions. Of particular interest here, is the Fourier transform of a function. Roughly speaking, to multiply the fourier coefficients of f by a sequence $\{b_n\}$ is equivalent to convoluting f with a function whose fourier transform is b_n at λ_n. Since the fourier transform is sort of its own inverse, we may start either with the function or its transform. We define the fourier transform by

$$\hat{\varphi}(u) = \frac{1}{2\pi}\int_{-\infty}^{\infty} e^{-iut}\varphi(t)\,dt.$$

In most cases φ will have compact support and have enough smoothness to insure that the inversion

$$\varphi(t) = \int_{-\infty}^{\infty} e^{itu}\hat{\varphi}(u)\,du$$

is valid.

Lemma 4.7. Let $f \in AP(E)$ and φ be a complex value function such that $\hat{\varphi} \subset L_1(R)$ and the inversion theorem holds. Define

$$\sigma(x) = \int_{-\infty}^{\infty} f(x+t)\hat{\varphi}(t)\,dt.$$

Then $\sigma \in AP(E)$ and

$$\sigma(t) \sim \sum a(f,\lambda)\varphi(\lambda)e^{i\lambda t}.$$

Furthermore

$$T(\sigma,\epsilon) \supset T(f,\frac{\epsilon}{\|\hat{\varphi}\|_1}).$$

Proof: The function σ is well defined since the integral clearly exists. Now $\sigma(x+\tau) - \sigma(x) = \int_{-\infty}^{\infty} [f(x+\tau+t) - f(x+t)]\hat{\varphi}(t)dt$ so that $v_\sigma(\tau) \leq \|\hat{\varphi}\|_1 v_f(\tau)$ whence σ is in AP(E) and the last conclusion holds. To compute the fourier series of σ we note that

$$\frac{1}{T}\int_0^T \sigma(t)e^{-i\lambda x}dx = \int_{-\infty}^{\infty}(\frac{1}{T}\int_0^T f(x+t)e^{-i\lambda x}dx)\hat{\varphi}(t)dt$$

by Fubini's Theorem. We continue then to get

$$\frac{1}{T}\int_0^T \sigma(x)e^{-i\lambda x}dx = \int_{-\infty}^{\infty}(\frac{1}{T}\int_t^{t+T} f(u)e^{-i\lambda u}du)e^{i\lambda t}\hat{\varphi}(t)dt.$$

Since $\frac{1}{T}\int_t^{t+T} f(u)e^{-i\lambda u}du$ converges to $a(f,\lambda)$ as $T \to \infty$, even uniformly in t, we may apply Lebesgue's Dominated Convergence Theorem to get

$$a(\sigma,\lambda) = \int_{-\infty}^{\infty} a(f,\lambda)e^{i\lambda t}\hat{\varphi}(t)dt = a(f,\lambda)\varphi(\lambda).$$

We are interested, of course, in choosing φ so that σ has some interesting property connected with f. Three are discussed in the sequel.

6. <u>Functions with bounded exponents</u>. The simplest case that arises in which the function σ of Lemma 4.7 is connected with f is the case when $\sigma = f$. It is clearly necessary that $\varphi(\lambda) = 1$ at each exponent of f and it is very difficult for $\hat{\varphi}$ to be in L_1 unless the set of exponents is severely restricted. An interesting case is the case of bounded exponents.

<u>Theorem 4.8</u>. Let $f \in AP(E)$, such that $\exp(f) \subset [-M,M]$. Then f has an extension into the complex plane that is an entire function. Thus all derivatives of f are in $AP(E)$. Furthermore, there is a constant $C(M)$ such that $\|f'\| \leq C(M)\|f\|$.

<u>Proof</u>: If $\exp(f) \subset [-M,M]$, we let φ be a function which is equal to 1 on $[-M,M]$, vanishes outside some finite interval, and is in $C^{\infty}(R)$. Then it is easy to see that $\hat{\varphi} \in L_1(R)$, and $\hat{\varphi}$ has an extension into the complex plane as an entire function. Then the inversion theorem holds and we write

$$\sigma(x) = \int_{-\infty}^{\infty} f(x+t)\hat{\varphi}(t)\,dt.$$

By Lemma 4.7,

$$\sigma(x) \sim \sum a(f,\lambda)\varphi(\lambda)e^{i\lambda x} = \sum a(f,\lambda)\varphi^{i\lambda x} \sim f,$$

since $\varphi(\lambda) = 1$ at every point where $a(f,\lambda) \neq 0$. By the uniqueness theorem $\sigma = f$. Then

$$f(x) = \int_{-\infty}^{\infty} f(x+t)\hat{\phi}(t)\,dt = \int_{-\infty}^{\infty} f(u)\hat{\phi}(u-x)\,du.$$

From the second representation, it follows that f has an extension into the complex plane and

$$f'(x) = -\int_{-\infty}^{\infty} f(u)\hat{\phi}'(u-x)\,du$$

so that $\|f'\| \leq \|f\| \, \|\hat{\phi}'\|_1$. Note that $\hat{\phi}' \in L_1(R)$ since it is the fourier transform of a C^{∞} function with compact support. We take $C(M) = \|\hat{\phi}'\|_1$. Note again that it does not depend on f or $\exp(f)$ but only on M.

Theorem 4.8 is an extension of a well-known theorem of Bernstein about periodic trigonometric polynomials. In the periodic case, it can be shown that $C(M) = M$. Even in the almost periodic case this is the best possible constant but we will not prove it.

7. Functions with discrete exponents. The second application of the convolution idea is to consider the case when σ is a trigonometric polynomial. If ϕ has compact support then it is necessary that bounded sets have only a finite number of exponents.

Theorem 4.9. Suppose that $f \in AP(E)$ and $\exp(f)$ has no finite limit point. Then there exist ϕ_n such that

$\sigma_n(x) = \int_{-\infty}^{\infty} f(x+t)\hat{\varphi}_n(t)\,dt$ are trigonometric polynomials that converge to f. Furthermore we have the estimate $\|f - \sigma_n\| \leq \frac{MK}{n}$ for an absolute constant M, if f is Lipschitz with constant K.

Proof: Let φ be a function with the following properties: $\varphi(0) = 1$, $\varphi > 0$ on $(-1,1)$, $\varphi(t) = 0$ if $|t| \geq 1$, and φ has three continuous derivatives. Define $\varphi_n(t) = \varphi(\frac{t}{n})$, and

$\sigma_n(x) = \int_{-\infty}^{\infty} f(x+t)\hat{\varphi}_n(t)\,dt$. It is easy to verify that

$\hat{\varphi}_n(t) = n\hat{\varphi}(nt)$ and therefore $\|\hat{\varphi}_n\|_1 = \|\hat{\varphi}\|_1$ for all n. Also the inversion theorem holds and $\lim_{n\to\infty} \varphi_n(t) = 1$ for any t. Now a look at Lemma 4.7 shows that $\sigma_n \sim \sum a(f,\lambda)\varphi_n(\lambda)e^{i\lambda t}$ so the fourier series converges formally to the fourier series of f as $n \to \infty$. So any subsequence that converges, must converge to f. Also $T(\sigma_n,\epsilon) \supset T(f,\frac{\epsilon}{\|\hat{\varphi}\|_1})$ shows that the functions σ_n form a u.a.p. family. Since from every subsequence of σ_n we can extract a further subsequence that converges uniformly (Theorem 2.5) and this limit function is always f, the entire sequence must converge to f. Note that σ_n is a trigonometric polynomial.

To get the estimate on $\|f - \sigma_n\|$ when f is Lipschitz with constant K we note that by the inversion theorem

$1 = \varphi_n(0) = \int_{-\infty}^{\infty} \hat{\varphi}_n(u)\,du$. Consequently,

$$\sigma_n(x) - f(x) = \int_{-\infty}^{\infty} (f(x+t) - f(x))\hat{\varphi}_n(t)\,dt. \quad \text{Thus}$$

$$|\sigma_n(x) - f(x)| \leq \int_{-\infty}^{\infty} K|t|\,|\hat{\varphi}_n(t)|\,dt = K\int_{-\infty}^{\infty} |t|\,|n\hat{\varphi}(nt)|\,dt$$

$$= \frac{K}{n}\int_{-\infty}^{\infty} |s|\,|\hat{\varphi}(s)|\,ds.$$

Now $M = \int_{-\infty}^{\infty} |s|\,|\hat{\varphi}(s)|\,ds < \infty$ since φ has three continuous derivatives. This gives $\hat{\varphi}$ behavior like $\frac{1}{s^3}$ at ∞ and the integral is finite.

Corollary 4.10. Let $\mathfrak{F}$ be a family of functions in $AP(E)$, whose exponents all lie in a given countable set Λ with no finite limit point. If there is a K so that $|f(t) - f(s)| \leq K|t-s|$ for all $t,s \in R$ and $f \in \mathfrak{F}$, and if there is an M_1 such that $\|f\| \leq M_1$ for all $f \in \mathfrak{F}$, then $\mathfrak{F}$ is a u.a.p. family. Conversely, if $\mathfrak{F}$ is the family of all functions in AP such that $\|f\| \leq M_1$ and $|f(t) - f(s)| \leq K|t-s|$ with $\exp(f) \subset \Lambda$, then $\mathfrak{F}$ is u.a.p. only if Λ has no finite limit point.

Proof: We invoke Theorem 4.9 to get the estimate $\|f - \sigma_n(f)\| \leq \frac{KM}{n}$. This means, that given $\epsilon > 0$, we can find an n so that (order Λ by $|\lambda_1| \leq |\lambda_2| \cdots$)

$$\sigma_n(f,t) = \sum_{k=1}^{N} a(f,\lambda_k)\varphi_n(\lambda_k)e^{i\lambda_k t}$$

approximates f within ϵ, and the estimate n is independent of f! In fact, pick n so large that $\frac{KM}{n} < \epsilon$, and let N be such that $|\lambda_N| > n$. Then $\varphi_n(\lambda_k) = 0$ if $k \geq N$ and σ_n is as above. Now the collection $\{\sigma_n(f,t) : f \in \mathfrak{F}\}$ is a family of u.a.p. functions, for $\|\sigma_n(f,t)\| \leq \|f\| + \epsilon \leq M_1 + \epsilon$ so they are uniformly bounded. Furthermore,

$$|\sigma_n(f,t+\tau) - \sigma_n(f,t)| \leq \sum_{k=1}^{N} |a(f,\lambda_k)| \; |\varphi_n(\lambda_n)| \; |e^{i\lambda_k\tau} - 1|$$
$$\leq \|f\| \; \|\varphi\| \sum_{k=1}^{N} |e^{i\lambda_k\tau} - 1|.$$

whence $T(\sigma_n,\epsilon) \supset \{\tau| \; |e^{i\lambda_k\tau} - 1| < \frac{\epsilon}{N\|f\| \; \|\varphi\|} = \epsilon_1, k = 1,-,N\}$. This is relatively dense and contains an interval about 0. In fact it is $\bigcap_{k=1}^{N} T(v_k,\epsilon_1)$ where $v_k(t) = e^{i\lambda_k t}$. Let v be the translation function for the family $\{\sigma_n(f,t)\}$. Since $\|\sigma_n(f,t) - f\| < \epsilon$, $v_f(\epsilon) \leq v_{\sigma_n}(3\epsilon) \leq v(3\epsilon)$. This holds for all $f \in \mathfrak{F}$ so $\mathfrak{F}$ is a u.a.p. family. To prove the converse, we suppose $\mathfrak{F}$ is the family with $\|f\| \leq M$ and $|f(t) - f(s)| \leq K|t-s|$ and $\exp(f) \subset \Lambda$ where Λ has a finite limit point and is a u.a.p. family. We can consider $\mathfrak{F}$ to be a subset of the Banach Space $\mathfrak{B}$ of almost periodic functions with exponents in Λ. Note that $\mathfrak{B}$ is infinite dimensional since $\{e^{i\lambda t} : \lambda \in \Lambda\}$ are independent functions in $\mathfrak{B}$. Note also that $\mathfrak{F}$ is closed in $\mathfrak{B}$, so that being u.a.p., it is compact. Let $\Lambda \cap [-N,N] = \Lambda_1$ be infinite.

If $\varphi \in \mathfrak{B}$ such that $\|\varphi\| \leq M_2$ and $\exp(\varphi) \subset \Lambda_1$ then φ' exists and is almost periodic and $\|\varphi'\| \leq M_2 N$ by Theorem 4.8. Choose M_2 so small that $M_2 N \leq K$ and $M_2 \leq M_1$. Then $S = \{\varphi \mid \varphi \in \mathfrak{B}, \|\varphi\| \leq M_2, \exp(\varphi) \subset \Lambda_1\}$ is a subset of $\mathfrak{J}$. Furthermore, it is clearly closed in $\mathfrak{J}$, so that in fact it is compact in $\mathfrak{B}$. On the other hand, S is a ball in the infinite-dimensional subspace $\mathfrak{B}_1$ of $\mathfrak{B}$ which are those elements of $\mathfrak{B}$ with exponents in Λ_1. Being compact in $\mathfrak{B}$, S is also compact in $\mathfrak{B}_1$. This is impossible. No ball in an infinite dimensional Banach space is compact.

8. Functions with exponents bounded away from zero. What we are interested in here is the theorem that if $\exp(f) \geq \lambda > 0$ then $\int_0^x f$ is almost periodic. Our approach is to prove the theorem for trigonometric polynomials and use the Approximation Theorem to extend the result. What we shall need is the fact that when $\{f_n'\}$ converges then so does $\{f_n\}$ converge. In our case $\{f_n'\}$ will be a sequence of polynomials converging to f'. This will be accomplished by proving a companion to the inequality of Theorem 4.8, namely that the reverse inequality $\|f\| \leq D(M)\|f'\|$ holds if $|\exp(f)| \geq M$.

Lemma 4.11. Let f be a trigonometric polynomial $f(t) = \sum_{n=1}^{N} a_n e^{i\lambda_n t}$ with $|\lambda_n| \geq M$. Let $g(t) = \sum_{n=1}^{N} \frac{a_n}{i\lambda_n} e^{i\lambda_n t}$. There exists a constant D independent of f, g and M such that $\|g\| \leq DM^{-1}\|f\|$.

Proof: Note that $g' = f$ and g is the unique primitive of f with mean value zero. We employ the idea of Lemma 4.7. Take first the case when $M = 1$ and define $\varphi(t) = it$ $|t| \leq 1$ and $(-it)^{-1}$ if $|t| \geq 1$. We first get an estimate for the fourier transform of the form $|\hat{\varphi}(u)| \leq C(1+u^2)^{-2}$ for $u \neq 0$. If $|t| \geq 1$ then $t^3 \geq \frac{1}{2}(1+t^2)$, consequently,

$$2\pi\hat{\varphi}(u) = \int_{-\infty}^{\infty} e^{-iut}\varphi(t)\,dt = \frac{1}{iu}\int_{-\infty}^{\infty} e^{-iut}\varphi'(t)\,dt$$

$$= \frac{1}{u}\int_{-1}^{1} e^{-iut}dt - \frac{ie^{-iut}}{(tu)^2}\Bigg|_{1}^{-1} - \frac{2i}{u^2}\int_{|t|\geq 1} \frac{e^{-iut}}{t^3}\,dt$$

$$= \frac{2i}{u^2}(e^{-iu} - e^{iu}) - \frac{2i}{u^2}\int_{|t|\geq 1} \frac{e^{-iut}}{t^3}\,dt$$

$$\leq 2u^{-2} + 2u^{-2}\int_{|t|\geq 1} \frac{dt}{\frac{1}{2}(1+t^2)} = Cu^{-2} \leq \frac{2C}{1+u^2} \quad \text{if } |u| \geq 1.$$

For $|u| \leq 1$ and $u \neq 0$

$$|2\pi\hat{\varphi}(u)| = \left|\int_{-1}^{1} i\,t\,e^{-iut}dt + i\int_{|t|\geq 1} \frac{e^{-iut}}{t}\,dt\right|$$

$$\leq 2 + \left|\int_{1}^{\infty} \frac{\sin ut}{t}\,dt\right| = 2 + \pi \leq \frac{2(2+\pi)}{1+u^2}.$$

It now follows that $\hat{\varphi} \in L_1(R)$ and the inversion theorem holds in the sense of Cauchy principal value. Then we have

$$g(t) = -\sum_{n=1}^{N} a_n \varphi(\lambda_n) e^{i\lambda_n t} = -\sum_{n=1}^{N} a_n \int_{-\infty}^{\infty} \hat{\varphi}(u) e^{iu\lambda_n} du\, e^{i\lambda_n t}$$

$$= -\int_{-\infty}^{\infty} \Big(\sum_{n=1}^{N} a_n e^{i(u+t)\lambda_n} \Big) \hat{\varphi}(u)\, du = -\int_{-\infty}^{\infty} f(u+t) \hat{\varphi}(u)\, du$$

whence $\|g\| \leq \|\hat{\varphi}\|_1 \|f\|$. Note that $\|\hat{\varphi}\|_1$ does not depend on f, g, or the exponents, but is an absolute constant. To do the general case, suppose $|\lambda_n| \geq M$, consider $h(t) = f(\frac{t}{M})$ and $k(t) = Mg(\frac{t}{M})$. Then $h(t) = \sum_{n=1}^{N} a_n e^{i(\frac{\lambda_n}{M})t}$ and

$$k(t) = \sum_{n=1}^{N} \frac{a_n e^{i\frac{\lambda_n}{M}t}}{i(\frac{\lambda_n}{M})}$$ where the exponents $\mu_n = \frac{\lambda_n}{M}$, satisfy

$|\mu_n| \geq 1$ and hence $\|k\| \leq \|\hat{\varphi}\|_1 \|h\|$ so that
$\|g\| = \|k\| M^{-1} \leq \|\hat{\varphi}\|_1 M^{-1} \|f\|$.

<u>Theorem 4.12</u>. Let $f \in AP(E)$ such that $f(t) \sim \sum a_n e^{i\lambda_n t}$ with $|\lambda_n| \geq M > 0$. Then $\int_0^t f(s)\,ds$ is in $AP(E)$ and if g is the integral of f with $a(g,0) = 0$, then $\|g\| \leq DM^{-1}\|f\|$. D is an absolute constant.

<u>Proof</u>: We let p_n be a sequence of trigonometric polynomials with $\exp(p_n) \subset \exp(f)$ which converge uniformly to f. If

$$p_n(t) = \sum_{k=1}^{N} b_k^{(n)} e^{i\lambda_k t}$$ define

$q_n(t) = \sum_{k=1}^{N} \frac{b_k^{(n)}}{i\lambda_k} e^{i\lambda_k t}$. Note that under this general correspondence, $(p_n - p_m)(t)$ and $(q_n - q_m)(t)$ are related in the same way so that we may apply Lemma 4.11, to get $\|q_n - q_m\| \leq D(M)\|p_n - p_m\|$. Since $\|p_n - p_m\| \to 0$ as $n, m \to \infty$, it follows that the sequence $\{q_n\}$ converges uniformly. Let g be its limit. Then from $\|q_n\| \leq D(M)\|p_n\|$ we deduce that $\|g\| \leq D(M)\|f\|$. Since $p_n \to f$ it follows that $b_k^{(n)} \to a(f,\lambda_k)$ as $n \to \infty$ so that

$g \sim \sum \frac{a(f,\lambda_n) e^{i\lambda_n t}}{i\lambda_n}$ and g is a formal integral of f with mean value zero. Since $q_n' = p_n$ for all n, a standard advanced calculus theorem gives $g' = f$ at every point of R. This completes the proof.

Theorem 4.12 is, of course, a theorem about the solutions of the differential equation $y' = f$. We will have occasion in the next chapter to generalize this result considerably. But the generalization is a very soft proof compared to this one. However, it will appeal directly to this result.

9. Notes. The proof of Kronecker's Theorem and Theorems 4.1 and 4.2 can be found in Bohr, Collected Works. The various statements of Theorem 4.5 can be found in Favard, Lecons sur les fonctions presque-periodiques, Gautier-Villars, Paris, 1933. We have added some that are appropriate to later discussions.

The trick of finding new almost periodic functions by convoluting with other functions is borrowed from the periodic case and has been exploited by many authors. It is extensively used in Levitan, Pocti-periodiceski funkcii, Gosudarstv Izdat Tehn.-Teor. Lit. Moscow, 1953. Theorem 4.12 may also be found in that book.

The particular result of Theorem 4.9 is in Bredhina, Some problems in summation of Fourier series of almost periodic functions, Amer. Math. Soc. Transl.(2) 26(1963), 253-261, and that of Theorem 4.8 is a remark of Bochner, Properties of Fourier Series of almost periodic functions, Proc. London Math. Soc., 26 (1927), 433-452. Theorem 4.11 is due to Bohr but the proof we give is in Coppel, Amost Periodic properties of ordinary differential equations, Ann. Mat. Pura Appl. 76(1967), 27-50.

More on module containment can be found in Cartwright, Comparison Theorems for Almost Periodic Functions, J. London Math. Soc. (2) 1(1969), 11-19.

Chapter 5

Linear Constant Coefficient Equations

1. Introduction. We are now in a position to consider the class of linear differential equations with constant coefficients. Roughly speaking there are two sorts of things that work. First, boundedness of solutions is equivalent to their almost periodicity. Secondly, since in many cases the solutions can be given explicitedly, one can give a direct verification. The plan of attack is essentially to reduce all cases to the trivial one $y' = f$. Direct proofs are available for this case, but we prefer to use an idea that will re-occur in the more complicated case of nonconstant coefficients.

2. Existence of minimum norm soltuions. We consider the following idea. If one considers the differential equation $y' = Ay + f$ where A is a matrix and f and y are vectors, it is easy to verify that the set of solutions is a convex set. In fact the bounded solutions are a convex set that does not contain 0 if $f \neq 0$. Such sets should have an element closest to zero, i.e. with minimum norm. This solution plays a distinguished role in our further discussion. Inasmuch as we need to look at this problem on all of R, it is not clear that such a minimum in fact exists.

Theorem 5.1. Let the set of bounded solutions of a differential equation $x' = f(x,t)$ be non-empty. If f is defined and bounded on $S \times R$ where S is a sphere containing the origin and a bounded solution, then there is a solution of the equation with minimum norm.

Proof: If x_o is one bounded solution we take K to be any compact set in E^n containing the range of x_o. We let $\lambda = \inf\{\|x\| : x(t) \in K$ for all $t \in R$ and x is a solution of $x' = f(x,t)\}$. Let $\lambda_n = \inf\{ \sup_{|t|\leq n} |x(t)| : x(t) \in K$ for all t and x is a solution of $x' = f(x,t)\}$. It is clear that $\lambda_n \leq \lambda_{n+1}$, and $\lambda = \lim_{n\to\infty} \lambda_n$. Let x_n be a solution such that $\sup_{|t|\leq n} |x_n(t)| \leq \lambda_n + 1/n$. According to Theorem 2.4 we may find a subsequence n_k of the integers so that $x_{n_k}(t) \to y(t)$ uniformly on all compact subsets of R and y is a solution of $x' = f(x,t)$. Now for a fixed t, we have for large k that $|x_{n_k}(t)| \leq \lambda_{n_k} + 1/n_k$ so upon taking limits we have $|y(t)| \leq \lambda$. Thus $\|y\| = \lambda$ since $y(t) \in K$ for all t.

3. The differential equation $y' = f(t)$. We now can look at the simplest differential equation $x' = f(t)$, where f is in $AP(E^n)$. One notes that either all solutions, or no solutions are in $AP(E^n)$, and the same can be said about boundedness. The two properties turn out to be equivalent. This should not be surprising for this is already true in the periodic case. To see this, note that by consideration of the fourier series, the solution has the same period as f, if it is periodic. But

$\int_0^t f(s)ds$ is periodic with same period if and only if $\int_0^T f(s)ds = M(f) = 0$ (T the period). If $\int_0^T f(s)ds \neq 0$ then $\int_0^{nT} f(s)ds = n\int_0^T f(s)ds$ is unbounded.

<u>Theorem 5.2</u>. Let $f(t) \in AP(E^n)$ and $x' = f(t)$ for all t. Then $x \in AP(E^n)$ if and only if it is bounded on R. If x is almost periodic, then $\exp(x) \subset \exp(f) \cup \{0\}$.

<u>Proof</u>: If $x \in AP(E^n)$ then it is bounded so it is the converse that we must prove. Let x be a bounded solution. According to Theorem 5.1 there is a solution with minimum norm. Furthermore, this solution is the only one giving the minimum, since any two solutions differ by a constant. This statement is true for all equations of the form $x' = g$, in particular for those with $g = T_\alpha f$. Let x_o be the minimizing solution for $x' = f$, and $g = T_\alpha f$ uniformly, then the claim is that there is an $\alpha' \subset \alpha$ so that $T_{\alpha'} x_o$ is the minimizing solution for $x' = T_\alpha f$. By Theorem 2.4 we can find $\alpha' \subset \alpha$ so that $T_{\alpha'} x_o = y_o$ exists uniformly on compact sets and y_o is a solution of $x' = T_\alpha f$. Repeat this argument with $-\alpha'$ to get $z_o = T_{-\alpha''} y_o$ a solution of $x' = T_{-\alpha''} T_\alpha f = f$ with $\alpha'' \subset \alpha'$. It is clear that $\|z_o\| \le \|y_o\| \le \|x_o\|$. Hence $\|x_o\| = \|z_o\|$ and $\|y_o\| = \|x_o\| = \lambda$. Thus the minimizing norm, λ is a constant for all equations of the form $T_\alpha f$. In particular, the minimizing solution always can be obtained by translation of one solution x_o. We can now prove the almost periodicity of x_o. For let α' and β' be given

sequences. Find common subsequence in turn $\alpha \subset \alpha'$ $\beta \subset \beta'$ so that $T_{\alpha+\beta}f = T_\alpha T_\beta f$ by the almost periodicity of f, then by Theorem 2.4 take more subsequences so that $T_\beta x_o$, $T_\alpha T_\beta x_o$, and $T_{\alpha+\beta}x_o$ exists uniformly on compact sets. By the above $T_\alpha T_\beta x_o$ and $T_{\alpha+\beta}x_o$ are minimizing solutions of the same equation $x' = T_{\alpha+\beta}f$ so are the same by uniquess, i.e. $T_{\alpha+\beta}x_o = T_\alpha T_\beta x_o$, and x_o is almost periodic. Now the original bounded solution $x(t) = x_o(t) + a$ for some constant a, so is almost periodic.

4. <u>The equation $x' = Ax$</u>. The next simple differential equation that we want to handle is the homogeneous linear equation $x' = Ax$, A a matrix. It is well known that the vector solutions for constant A are given by $e^{At}x_o$, and that elements of e^{At} are of the form $\sum_{i=1}^{k} p_i(t)e^{\lambda_i t}$ where the λ_i are the eigenvalues of A and the p_i are polynomials whose degrees do not exceed the multiplicity of λ_i. In the search for solutions that are almost periodic, we are interested in bounded solutions. It is thus clear that if $\text{Re}(\lambda_i) \neq 0$ for all i, then no solution except 0 is bounded. On the other hand if some λ_i is pure imaginary, and the corresponding polynomial p_i is constant, then there are bounded solutions and they in fact correspond to almost periodic solutions which are trigonometric polynomials. We summarize.

<u>Theorem 5.3</u>. The bounded solutions of $x' = Ax$, A constant, are precisely the almost periodic solutions.

5. <u>The $n^{\underline{th}}$ order scalar equations</u>. The standard way to handle the $n^{\underline{th}}$ order scalar equation

$$y^{(n)} + a_1(t)y^{(n-1)} + \cdots + a_n(t)y = g(t)$$

is to make the change of variables $x_1 = y, x_2 = y', \ldots, x_n = y^{(n-1)}$ to get the system $x' = A(t)x + f(t)$ where $x = (x_1,\ldots,x_n)^T$ $f(t) = (0,\ldots,0,g(t))^T$ and

$$A(t) = \begin{pmatrix} 0 & 1 & & 0 \\ & \ddots & \ddots & \\ & & \ddots & \ddots \\ 0 & & 0 & 1 \\ -a_n(t) & & -a_2 & -a_1(t) \end{pmatrix}.$$

We wish to do this also. For the case when $g(t) = 0$, so that $f(t) = 0$, and when A is a constant, we would like to have Theorem 5.3 cover the scalar equation. Now it appears that the hypothesis of Theorem 5.3 is that the vector x is bounded. In the scalar case, this is the statement that $y, y', \ldots, y^{(n-1)}$ are all bounded. But reversing the procedure, one calls y a bounded solution of the scalar equation even if one of the derivatives may be unbounded. In the constant coefficient case, this probably never happens, but certainly in the non-constant case it might happen that, for example, $\cos(t^2)$ is a solution which is bounded, but has unbounded derivative. It turns out that if $\cos(t^2)$ is a solution of a linear differential equation then it must have unbounded coefficients. This is the content of the next theorem.

<u>Theorem 5.4</u>. Let

$$Ly = f(t) \qquad\qquad L = \sum_{i=0}^{n-1} a_i(t) \frac{d^i}{dt^i} + \frac{d^n}{dt^n}$$

be a linear differential equation where the coefficients a_i and f are bounded on all of R. If y is a solution bounded on R, then $y',\ldots,y^{(n)}$ are bounded on R as well.

<u>Proof</u>: We will prove this theorem by using the fact (Lemma 5.6 below) that if I is a sufficiently large interval, then there is a constant A so that for any n times differential function with $\|y''\| \geq \|y\|$, one has

$\|y^{(k)}\| \leq A\|y\|^{1-k/n}\|y^{(n)}\|^{k/n}$, $k = 1,\ldots,n-1$ and A does not depend on I. Here the norms are the maximum over the interval I. The condition that $\|y''\| > \|y\|$ is easy to achieve by a change of scale $t \to at$ when $\|y\|$ is uniformly bounded. Select an interval I of large length and let norms be over this interval, then

$$\|y^{(n)}\| \leq \sum_{i=1}^{n-1} \|a_i\| \, \|y^{(i)}\| + \|a_n\| \, \|y\| + \|f\|.$$

Let B be a number so that $f,y,a_1,\ldots,a_{n-1}$ all have norms less than B on all of R. We then get

$$\|y^{(n)}\| \le B \sum_{k=1}^{n-1} A\, B^{1-k/n} \|y^{(n)}\|^{k/n} + 2B$$
$$\le C \sum_{k=0}^{n-1} \|y^{(n)}\|^{k/n}$$

where C is a sufficiently large number depending only on the absolute constants A and B. Now if $\|y^{(n)}\|_I \le 1$ for all intervals I then we are done since $\|y^{(k)}\| \le A\|y\|^{1-k/n} \le AB$ on R. On the other hand, if there is an interval J over which $\|y^{(n)}\| > 1$ then it is also larger than 1 if $I \supset J$. Restrict I this way and we have $\sum_{k=0}^{n-1} \|y^{(n)}\|^{k/n} \le n\|y^{(n)}\|^{\frac{n-1}{n}}$ so that $\|y^{(n)}\| \le Cn\|y^{(n)}\|^{\frac{n-1}{n}}$. Thus $\|y^{(n)}\| \le (Cn)^n$. This constant is independent of I and so $\|y^{(n)}\|$ is uniformly bounded. So are the lower order derivatives.

6. The Differential Inequalities. To complete the proof of Theorem 5.4, we need to prove two lemmas. The first is the case $n = 2$ and serves as the basis of an induction argument for the general case.

Lemma 5.5. If $I = [a,b]$ is an interval such that $b - a \ge 2$ and y is twice differentiable in I with $\|y''\| > \|y\|$, then $\|y'\| \le 2\|y\|^{1/2}\|y''\|^{1/2}$, (norms on I).

Proof: One writes for $t \in (c,d)$ $a \le c < d \le b$

$$y(c) = y(t) + y'(t)(c-t) + y''(\xi_1)\frac{(c-t)^2}{2} \quad \text{and}$$

$y(d) = y(t) + y'(t)(d - t) + y''(\xi_2)\dfrac{(d - t)^2}{2}$ for certain ξ_1 and ξ_2 in (a,b). Subtracting these two one has

$$(d - c)|y'(t)| \leq 2\|y\|_{[c,d]} + \|y''\|_{[c,d]}\frac{(c - t)^2 + (d - t)^2}{2}$$

$$\leq 2\|y\|_I + \|y''\|_I \frac{(d - c)^2}{2} \qquad \text{or}$$

$$|y'(t) \leq \frac{2}{d - c}\|y\|_I + (d - c)\frac{\|y''\|_I}{2} .$$

If $\|y'\| = |y'(t)|$ select c and d so that $t \in (c,d)$ and $d - c = 2\left(\dfrac{\|y\|_I}{\|y''\|_I}\right)^{1/2}$. This is possible since $b - a > 2$ and $\|y\|_I < \|y''\|_I$. One then has

$$\|y'\| = |y'(t)| \leq \frac{2}{d - c}\|y\|_I + \frac{d - c}{2}\|y''\|_I = 2\|y\|_I^{1/2}\|y''\|_I^{1/2}.$$

<u>Lemma 5.6</u>. There are constants A_n, depending only on n, such that, if $I = [a,b]$ is such that $b - a \geq 2$ and $\|y''\|_I \geq \|y\|_I$, then

$$\|y^{(k)}\| \leq A_n\|y\|^{1-k/n}\|y^{(n)}\|^{k/n} , \qquad k = 1,\ldots,n-1 ,$$

whenever y is n times differentiable and the norms are over I.

<u>Proof</u>: We do an induction with Lemma 5.5 serving as the first step. If the constant A_{n-1} exists we proceed to the $n^{\underline{th}}$ case. First for $k = 1$ we have $\|y'\| \leq A_2\|y\|^{1/2}\|y''\|^{1/2}$ by Lemma 5.5 and $\|y''\| \leq A_{n-1}\|y'\|^{1 - \frac{1}{n-1}}\|y^{(n)}\|^{\frac{1}{n-1}}$ by the induction hypothesis. Eliminating y'' from these one has

$\|y'\| \le A_2 A_{n-1}^{1/2}\|y\|^{1/2}\|y'\|^{\frac{n-2}{2(n-1)}}\|y^{(n)}\|^{\frac{1}{2(n-1)}}$. Solving this inequality for $\|y'\|$ one has $\|y'\| \le A_2^{2(\frac{n-1}{n})} A_{n-1}^{\frac{n-1}{n}}\|y\|^{1-1/n}\|y^{(n)}\|^{1/n}$. If $2 \le k \le n-1$, then we write $\|y^{(k)}\| \le A_{n-1}\|y'\|^{1-\frac{(k-1)}{n-1}}\|y^{(n)}\|^{\frac{k-1}{n-1}}$ and use the above estimate on $\|y'\|$ to get

$\|y^{(k)}\| \le A_2^{2\frac{n-k}{n-1}} A_{n-1}^{\frac{2n-k}{n}}\|y\|^{1-k/n}\|y^{(n)}\|^{k/n}$. Take $A_n \ge A_2^{2\frac{n}{n-1}} A_{n-1}^2$.

7. Almost Periodic Solutions of $x' = Ax$. There is one more salient fact that we need about almost periodic solutions of $x' = Ax$ which is just as easy to prove for almost periodic A as for constant A.

Theorem 5.7. If $A(t)$ is an almost periodic matrix and x is an almost periodic solution to $x' = A(t)x$, then $\inf_{t \in R}|x(t)| > 0$ or else $x(t) \equiv 0$.

Proof: If $\inf|x(t)| = 0$, let $|x(\alpha_n')| \to 0$. Find $\alpha \subset \alpha'$ such that $T_\alpha A = B$, $T_{-\alpha}B = A$, $T_\alpha x = y$, and $T_{-\alpha}y = x$ exist uniformly, see Theorem 1.17. Then $y' = By$ and $y(0) = 0$. Thus $y \equiv 0$ and $x = T_{-\alpha}y \equiv 0$.

8. The general linear equation. We are now in a position to consider the equation

$$x' = Ax + f$$

where $f \in AP(E^n)$ and A is a constant matrix. We will show how to use two distinct methods to handle this equation. Each has its virtues. Our first method is a direct generalization of the proof of Theorem 5.2.

Theorem 5.8. A solution x of $x' = Ax + f$ is almost periodic if and only if it is bounded. If there is a bounded solution, then there is an almost periodic solution x_1, such that $\mathrm{Mod}(x_1) = \mathrm{mod}(f)$.

Proof: We assume that there is a bounded solution x_o and thus there is a solution $x(f)$ of minimum norm. If $T_{\alpha'} f = g$ uniformly, then there is $\alpha \subset \alpha'$ so that $y = T_\alpha x_o$, uniformly on compact sets, is a bounded solution of $x' = Ax + g$. Thus every equation of the form $x' = Ax + T_\alpha f$ has a solution of minimum norm $x(T_\alpha f)$. Say a typical equation of this form is $x' = Ax + g$. We now claim that $x(g)$ is unique. If x_1 and x_2 are distinct solutions of $x' = Ax + g$ such that $\|x_1\| = \|x_2\| = \|x(g)\|$ then $\frac{x_1 + x_2}{2}$ is a solution of $x' = Ax + g$ and $\frac{x_1 - x_2}{2}$ is a non-trivial solution of $x' = Ax$ so that $|\frac{x_1 - x_2}{2}(t)| \geq \rho > 0$ for all t. Now by the parallelogram law

$$|\frac{x_1 + x_2}{2}(t)|^2 + |\frac{x_1 - x_2}{2}(t)|^2 = \frac{|x_1(t)|^2 + |x_2(t)|^2}{2} \leq \|x(g)\|^2.$$

Consequently, $|\frac{x_1 + x_2}{2}(t)|^2 \leq \|x(g)\|^2 - \rho^2$ and $\|\frac{x_1 + x_2}{2}\| < \|x(g)\|$ contradicting the minimum property of $\|x(g)\|$.

Now we finish the proof as in Theorem 5.2. For any β it follows that $T_\alpha x(f) = x(T_\alpha f)$ and hence $T_{\alpha+\beta}x(f) = T_\alpha T_\beta x(f)$ since both minimize the norm over bounded solutions of $x' = Ax + T_{\alpha+\beta}f$. Thus $x_o(t) = x(f)(t) + z(t)$ where z is a bounded solution of $y' = Ay$ which is almost periodic by Theorem 5.3. This proves the almost periodicity of x_o. Note that $T_\alpha f = f$ uniformly implies that $T_{\alpha'}x(f) = x(f)$ for some $\alpha' \subset \alpha$. By Theorem 4.5 $\text{Mod}\, x(f) \subset \text{Mod}(f)$. But $f = x' - Ax$ shows the reverse.

9. The scalar equation. Although Theorem 5.8 is fairly definitive, we show how to get the theorem in a more constructive way. The idea is to look at the scalar equation and build the vector equation from a short sequence of scalar equations. We thus consider the scalar equation

$$x' = \alpha x + f(t) \tag{5.1}$$

where $f \in AP(E)$ and α is a complex number. We know that bounded solutions are almost periodic, but we can sometimes exhibit the solution and show more about the mapping from equation to solutions.

Theorem 5.9. If $\text{Re}(\alpha) \neq 0$, and $f \in AP(E)$, then there is a unique bounded solution $x(f)$ of (1) which is almost periodic. Also $\exp(x(f)) = \exp(f)$ and the mapping $f \to x(f)$ is a linear mapping from $AP(E)$ to itself with norm $|\text{Re}(\alpha)|^{-1}$.

Proof: If $\text{Re}(\alpha) \neq 0$, then there is at most one bounded solution. If $\text{Re}(\alpha) > 0$, then

$x(t) = -\int_t^\infty e^{\alpha(t-s)} f(s)\,ds$ is easily verified to be a solution with $\|x\| \leq \mathrm{Re}(\alpha)^{-1}\|f\|$. If $\mathrm{Re}(\alpha) < 0$ then take $x(t) = \int_{-\infty}^t e^{\alpha(t-s)} f(s)\,ds$. In this case $\|x\| \leq (-\mathrm{Re}(\alpha))^{-1}\|f\|$. In the first case, we may write $x(t) = -\int_0^\infty e^{-\alpha u} f(u+t)\,dt$ so that $|x(t+\tau) - x(t)| \leq |\mathrm{Re}(\alpha)|^{-1}\|f(t+\tau) - f(t)\|$ and hence x is almost periodic with $T(x,\epsilon) \supset T(f,\epsilon|\mathrm{Re}(\alpha)|)$. We could argue from Theorem 4.5, that $\mathrm{Mod}(x) \subset \mathrm{Mod}(f)$, and the reverse is true since $f = x' - \alpha x$. In fact more is true in this case. Since x' is almost periodic, $\exp(x') \subset \exp(x)$ so $\exp(f) \subset \exp(x)$. One can argue the equality from the fourier series. For if $x \sim a_0 + \sum_{n=1}^\infty a_n e^{i\lambda_n t}$, then $x' \sim \sum_{n=1}^\infty i\lambda_n a_n e^{i\lambda_n t}$. The differential equation then gives $i\lambda_n a_n - \alpha a_n = a(f,\lambda_n)$ $n \geq 1$ and $-\alpha a_0 = a(f,0)$. Since $i\lambda_n - \alpha \neq 0$ for all n we see that $a_n \neq 0$ if and only if $a(f,\lambda_n) \neq 0$.

Clearly the mapping $f \to x(f)$ is linear. The proof is complete.

Theorem 5.10. Suppose $\alpha = i\lambda$ for some real λ such that $|\lambda - \lambda_n| \geq d > 0$ for all $\lambda_n \in \exp(f)$. Then there exists a unique almost periodic $x(f)$ solution to (5.1) whose exponents are the same as the exponents of f. Furthermore, if $N_d \equiv \{f \mid f \in AP(E),\ |\lambda - \lambda_n| \geq d \text{ for all } \lambda_n \in \exp(f)\}$, then the mapping from N_d to itself, $f \to x(f)$, is a continuous linear map with norm less than Dd^{-1} where D is an absolute constant.

Proof: We look at the equation (5.1) and change variables by $y(t) = e^{-i\lambda t}x(t)$. Then $y'(t) = e^{-i\lambda t}f(t) \equiv g(t)$. Note that $\|g\| = \|f\|$, $\|x\| = \|y\|$, $\exp(y) = \exp(x) - \lambda$, and $\exp(g) = \exp(f) - \lambda$. If f satisfies the hypothesis then $\exp(g)$ are bounded away from 0 by a distance d. We apply Theorem 4.11 to get the unique integral y of g with $M(y) = 0$, y almost periodic, and $\|y\| \leq Dd^{-1}\|g\|$. Thus $\exp(y) = \exp(g)$ so that we now unravel our change of variables to get the theorem.

In order to get a common formulation of the previous two theorems, note that in both cases, $\text{dist}(\alpha, i\lambda_n) \geq \rho > 0$, $\lambda_n \in \exp(f)$, gives the existence of an almost periodic solution with the mapping from f to $x(f)$ having norm less than $D\rho^{-1}$.

Note that the formal solution cannot be almost periodic if $\sum \frac{|a(f,\lambda_n|^2}{|\alpha - i\lambda_n|^2} = +\infty$ so that some sort of condition is required. Certainly $\alpha \neq i\lambda_n$ is required. This is called a non-resonance condition.

10. The vector equation again. We again turn our attention to the vector equation

$$x' = Ax + f. \tag{5.2}$$

In order to reduce this to a succession of scalar equations, we note that for any matrix A, there exists a P so that

$PAP^{-1} = T$ is a lower triangular matrix. We then change variables by $y = Px$ to get

$$y' = Ty + g \tag{5.3}$$

where $g = Pf$.

Note that $\exp(y) \subset \exp(x)$ by $y = Px$ and the reverse follows since $x = P^{-1}y$. Also $\|x\| \, \|P^{-1}\|^{-1} \leq \|y\| \leq \|P\| \, \|x\|$ so that statements about norms on each equation can be transfered to the other.

Now the first equation in (5.3) is $y_1' = t_{11}y_1 + g_1$ and so is of the form of the scalar equation of the previous section. Here t_{11} is an eigenvalue of A. Thus the correct hypothesis is that $|\mu - i\lambda_n| \geq \rho > 0$ for μ an eigenvalue of A and $\lambda_n \in \exp(f)$. With this we get a unique y_1, almost periodic with $\exp(y_1) = \exp(g_1)$ and $\|y_1\| \leq D\rho^{-1}\|g_1\|$.

We now proceed to the second equation of (5.3) which is

$$y_2' = t_{22}y_2 + (g_2 + t_{21}y_1) = t_{22}y_2 + h_2$$

where $h_2 = g_2 + t_{21}y_1$ is almost periodic with $\exp(h_2) \subset \exp(f)$. With the above hypothesis we get a unique almost periodic y_2 with $\exp(y_2) = \exp(h_2)$ and $\|y_2\| \leq D\rho^{-1}\|h_2\| \leq$ $D\rho^{-1}(\|g_2\| + |t_{21}|\|y_1\|) \leq D\rho^{-1}(\|g\| + |t_{21}|D\rho^{-1}\|g\|)$ $= D\rho^{-1}\|g\|(1 + |t_{21}|D\rho^{-1})$. Proceeding in this way we get a solution y with $\exp(y) \subset \exp(f)$ and $\|y\| \leq \|g\|p(D\rho^{-1})$ where p is a polynomial of degree n with no constant term. The coefficients of p depend only on the matrix T so ultimately only on matrix A, in fact, only on the modulus of elements of A and P. Transfering this inequality back to the original equation, multiplies this polynomial by norms of P and P^{-1}. We can summarize the above description.

<u>Theorem 5.11</u>. Suppose that $|\mu - i\lambda_n| \geq \rho > 0$ for all eigenvalues μ of A and $\lambda_n \in \exp(f)$. Then there is a unique almost periodic solution $x(f)$ of (5.2) with $\exp(x(f)) = \exp(f)$. There exists a polynomial p of degree $\leq n$ with no constant term, depending only on the matrix A, and an absolute constant D, so that the mapping $f \to x(f)$ defined on $N_\rho = \{f \mid f \in AP(E^n)_1 |\mu - i\lambda_n| \geq \rho$, μ eigenvalue of A, $\lambda_n \in \exp(f)\}$ is a linear mapping with norm less than $p(D\rho^{-1})$.

<u>Proof</u>: The existence of $x(f)$ has been shown above together with the constant D and the polynomial p. We need to show the uniqueness and the linearity. If there are two solutions x_1 and x_2 of (5.2) with exponents in $\exp(f)$, then $(x_1 - x_2)$ is an almost periodic solution of the homogeneous equation with exponents in $\exp(f)$. Writing

$$x_1 \sim \sum_{n=1}^{\infty} a_n e^{i\lambda_n t} \qquad\qquad \lambda_n \in \exp(f)$$

and

$$x_2 \sim \sum_{n=1}^{\infty} b_n e^{i\lambda_n t}$$

we have

$$(x_1 - x_2) \sim \sum_{n=1}^{\infty} (a_n - b_n) e^{i\lambda_n t}$$

$$(x_1 - x_2) \sim \sum_{n=1}^{\infty} i\lambda_n (a_n - b_n) e^{i\lambda_n t}$$

From $(x_1 - x_2)' = A(x_1 - x_2)$ we must have $(i\lambda_n)(a_n - b_n) = A(a_n - b_n)$. Clearly $i\lambda_n$ is not an eigenvalue of A so $a_n - b_n = 0$ for all $\lambda_n \in \exp(f)$. So $x_1 = x_2$ by the uniqueness theorem. Linearity now follows from the uniqueness since if $x_i' = Ax_i + f_i$. Then $(\alpha x_1 + \beta x_2)' = A(\alpha x_1 + \beta x_2) + (\alpha f_1 + \beta f_2)$ and $\exp(\alpha x_1 + \beta x_2) \subset \exp(f)$. Thus $x(\alpha f_1 + \beta f_2) = \alpha x(f_1) + \beta x(f_2)$.

11. Norms of mappings. The virtue of the formulation of Theorem 5.11 is its explicit determination of the form of the norm of the mapping. The virtue of the formulation of Theorem 5.8 is that its proof is elegant and generalizes. In fact, the proof of Theorem 5.11 also generalizes in a different way so both these formulations seem worthwhile. The major defect of Theorem 5.8 is the lack of the norm of the mapping $f \to x(f)$ even when it is known that $Re(\lambda) \neq 0$ for all eigenvalues λ of A. In this case, it is known that there exist projections P_1 and P_2 so that $P_1 + P_2 = I$

$$|e^{At}P_1| \leq k_1 e^{-\sigma_1 t} \quad t \geq 0 \quad \text{and}$$

$$|e^{At}P_2| \leq k_2 e^{\sigma_2 t} \quad t \leq 0 \quad \text{where}$$

σ_1 and σ_2 are positive numbers. If λ is an eigenvalue of A, and $Re(\lambda) < 0$ then $Re(\lambda) \leq -\sigma_1$ and if $Re(\lambda) > 0$ then $Re(\lambda) \geq \sigma_2$.

The function

$$x(t) = \int_{-\infty}^{t} e^{A(t-s)} P_1 f(s)\,ds - \int_{t}^{\infty} e^{A(t-s)} P_2 f(s)\,ds$$

is the unique bounded solution which is almost periodic (same proof as Theorem 5.9) and $\|x\| \leq (\frac{K_1}{\sigma_1} + \frac{K_2}{\sigma_2})\|f\|$. This gives the same sort of estimate as Theorem 5.11 in the case of all eigenvalues with non-zero real part.

It is possible to prove that the mapping $f \to x(f)$ is continuous without getting explicit estimates.

<u>Theorem 5.12</u>. Let Λ be a fixed subset of R, and consider $C(\Lambda) \equiv \{f \mid f \in AP(E^n),\ \exp(f) \subset \Lambda\}$. Suppose that for each $f \in C(\Lambda)$ there exists a unique $x(f) \in C(\Lambda)$ which solves $x' = Ax + f$. Then the mapping $f \to x(f)$ is a continuous linear operator.

<u>Proof</u>: Linearity follows from uniqueness. We need to show continuity. Consider the two spaces B_1 and B_2 where the set of points of B_1 is $C(\Lambda)$ provided with uniform norm, and $B_2 \equiv \{f \mid f$ and f' are in $C(\Lambda)\}$ and the norm is given by $\|f\|_2 = \|f\| + \|f'\|$. By Theorem 3.7, B_1 and B_2 are complete, hence Banach spaces. We say 0 has a null set of exponents so $0 \in B_1$ and $0 \in B_2$. Consider the mapping $T : x \to f$ defined by $f = x' - Ax$, where T has domain B_2. By hypothesis, T maps B_2 onto B_1 and is 1 to 1 on B_1. Furthermore $\|Tx\|_1 = \|f\|_1 = \|x' - Ax\| \leq \|x'\| + \|Ax\| \leq \|x'\| + \|A\|\ \|x\|$ $\leq \max(1,\|A\|)(\|x'\| + \|x\|) = \max(1,\|A\|)\|x\|_2$. Thus T is continuous.

By a theorem of Banach T^{-1} is continuous. That is, there is a constant C so that $\|x\|_2 \leq C\|f\|$. But $\|x\| \leq \|x\|_2$ so $\|x\| \leq C\|f\|$ as was to be shown.

12. Notes. Theorem 5.2 is due to Bohr and is patterned after one of Bohl who was considering finite sums of periodic functions.

The minimum norm idea is found in Favard, Lecons sur les fonctions presque-periodiques, Gauthier-Villars, Paris, 1933. We discuss several nonlinear versions of this idea later.

Theorem 5.4 is found in Landau, Über einen Satz von Herrn Esclangon, Math. Ann. 102(1929), 177-188, while Lemma 5.6 for $n = 2$ is in Landau and also in Hadamard, C. R. Soc. Math. France (1914), 68-72. The generalization to all n is due to Hardy and Littlewood. Inequalities of the type found in Lemma 5.6 with $I = R$ are called Kolmogorov Inequalities. The best possible constants are known. For a proof using almost periodic functions, see Bang, Une ineqalite de Kolmogoroff et les Fonctions Presque-periodiques, Kgl. Danske Videnskabens Selskab, Mat. Fysk Meddelser XIX(1941), 1-18.

Bohr and Neugebauer, Über lineare Differentialgleichungen mit konstanten Koeffizienten und fastperiodischer rechter Seite. Nachr. Ges. Wiss. Gottingen, Math.-Phys. Klasse(1926), 8-22, proved Theorem 5.8. The particular proof we give uses the Favard minimum norm idea and is simplified greatly by the use of Bochner's pointwise version of almost periodicity.

Theorem 5.10 is in Coppel, Almost periodic properties of ordinary differential equations, Ann. Mat. pura Appl., 76(1967),

27-50.

Theorem 5.7 also has an almost automorphic version. If x is almost automorphic then the conclusion is true with exactly the same proof. As a matter of fact, all of the theorems in this chapter have almost automorphic versions and the proofs hardly need a change from the almost periodic case.

Bochner, A new approach to almost periodicity, Proc. Nat. Acad. Sci. USA, 48(1962), 2039-2043, generalizes Theorem 5.8 to include terms of the form $x^{(k)}(t-\tau_k)$. These results require that the derivatives be bounded and uniformly continuous. It is possible to add terms which are convolutions with L_1 kernels, See Doss, On the almost periodic solutions of a class of integro-differential-difference equations, Ann. Math. (2), 81(1965), 117-123.

The fact that in Theorems 5.2 and 5.3 we must introduce equations other than the one being studied, namely the equations in the hull, is a plague that will remain with us in all the later chapters.

The history of Theorem 5.2 is quite interesting. Doss, On bounded functions with almost periodic differences, Proc. Amer. Math. Soc. 12(1961), 488-489, observes that

$F(x) = \int_0^x f(t)\,dt$ for $f \in AP(C)$ has the property (easily proved)

that $F(x+h) - F(x)$ is bounded and almost periodic for any h. Now this hypothesis makes sense without integration. So he poses the question; If F is bounded and all differences $F(x+h) - f(x)$ are almost periodic, is F almost periodic? This question makes sense on a group. The answer is in the affirmative. Bochner considered the question where only one non-trivial difference is almost periodic. Bochner's theorem [79], gives the answer even for the almost automorphic case, F must be uniformly continuous for this to be true. Further papers along this line are Berg, On functions with almost periodic and almost automorphic first differences, J. Phys. Mech. 10(1969), 239-246. Terhas, Almost automorphic functions on a topological group, Ph.D. dissertation, University of Illinois, 1970. Hurt and Schaeffer, Critical Linear difference equation: A study in pathology, J. Math. Anal. Appl. 33(1971), 408-424.

Chapter 6

Linear almost periodic equations

1. Introduction. The results for constant coefficient equations are quite definitive. As might be expected, for non-constant coefficients, the results are not as complete. Explicit solutions are not available and limiting procedures introduce equations different from the one under study.

2. A Counterexample. The first natural question that arises for non-constant coefficients is whether Theorem 5.3 is true as stated. Unfortunately, the answer is no.

Example 6.1. There is an almost periodic function g so that $\exp \int_0^t g$ is bounded, but not almost periodic. In order to do the construction, we will exhibit an almost periodic function g, so that $\int_0^t g(s)\,ds \leq 0$ for all t and such that $\int_0^{t_n} g(s)\,ds \to -\infty$ for a certain sequence t_n. It follows that $e^{\int_0^t g}$ is a bounded solution of $y' = gy$ with 0 infimum. By Theorem 5.7, $e^{\int_0^t g}$ is not almost periodic. To construct g let $n \geq 2$ and

$$f_n(x) = \begin{cases} -\dfrac{n}{2^{n-1}-1} & 1 \leq x \leq 2^{n-1}-1 \\ 0 & x = 0,\ 2^{n-1} \end{cases}$$

and linear in between these values on $[0,2^{n-1}]$. Then extend f_n to be odd and periodic of period 2^n. The salient properties of f_n are

$$\text{(i)} \int_0^{2^n} f_n(t)\,dt = 0, \qquad \text{(ii)} \int_0^{2^{n-1}} f_n(t)\,dt = -n,$$

$$\text{(iii)} \|f_n\| = \frac{n}{2^{n-1}-1}, \quad \text{and} \quad \text{(iv)} \int_0^t f_n(s)\,ds \leq 0 \quad \text{for} \quad t \geq 0.$$

Since $\sum \|f_n\| < \infty$, the function $g(t) = \sum_{n=2}^{\infty} f_n(t)$ is almost periodic. Furthermore $\int_0^t g(s)\,ds = \sum_{n=2}^{\infty} \int_0^t f_n(s)\,ds \leq 0$ for $t \geq 0$ and since all f_n are odd, g is odd so its integral is even and this holds for all $t \in R$. Also $\int_0^{2^{n-1}} g(s)\,ds \leq \int_0^{2^{n-1}} f_n(s)\,ds = -n$ since $\int_0^{2^{n-1}} f_j(s)\,ds \leq 0$ if $j \neq n$. Take $t_n = 2^{n-1}$.

3. Bounded solutions. In the constant coefficient case where explicit solutions are available, the boundedness of solutions on all of R is a quite clear phenomena. For the case of non-constant coefficients, indirect methods, for example, a Lypanov function approach, often give results of boundedness in the future. For almost periodic equations, this is enough to ensure the existence of a solution bounded on all of R. Later we will give an even stronger version of this statement, but for the moment we will prove this weaker statement.

Theorem 6.2. Let $x' = f(x,t)$ have a solution φ which is bounded on $[t_o,\infty)$. If $f(x,t)$ is almost periodic in t uniformly for $x \subset K = \overline{\{\varphi(t) : t \geq t_o\}}$, then there is a solution of the equation on all of R with values in K.

Proof: There exists an increasing sequence α'_n such that $\lim_n \alpha'_n = +\infty$ and $T_{\alpha'} f(x,t) = f(x,t)$ uniformly on $K \times R$. In fact, we may take $\alpha'_n \in T(f,\frac{1}{n}) \cap [\alpha'_{n-1} + 1,\infty)$, $\alpha'_o = 0$. Now consider $\varphi_n(t) = \varphi(t+\alpha'_n)$. Given any interval (a,∞), for n sufficiently large, $\{\varphi_n\}$ is defined on (a,∞) and is a solution of $x' = f(x,t+\alpha'_n)$ there. Furthermore, this sequence is uniformly bounded and equi-uniformly continuous on (a,∞). By taking, a as a sequence going to $-\infty$ and using the diagonalization argument as in Theorem 2.4, we have $\alpha \subset \alpha'$ so that $T_\alpha \varphi = \psi$ exists uniformly on compact subsets of R, and converges to a solution of $T_\alpha f(x,t) = f(x,t)$. Now ψ is defined on all of R and its values are in K.

4. Favard's Theorem. We will now discuss the linear almost periodic system

$$x' = A(t)x + f(t) \tag{6.1}$$

and its associated homogeneous equation

$$x' = A(t)x \tag{6.2}$$

where $A(t)$ is an almost periodic matrix and $f(t)$ is an almost periodic vector, each with complex valued components.

If $B \in H(A)$ we say that $x' = B(t)x$ is an equation in the hull of (6.2). We will want to apply the proof of Theorem 5.8 to equation (6.1). That proof also introduces the equations in the hull of (6.2) as well as those of the form $x' = B(t)x + g(t)$ where $T_\alpha A = B$ and $T_\alpha f = g$. This will be a typical equation in the hull of (6.1). To distinguish these, we call the latter an equation in the hull of (6.1) and the former an equation in the homogeneous hull of (6.1).

Theorem 6.3. Suppose that every non-trivial solution x of an equation in the homogeneous hull of (6.1) that is bounded on R satisfies $\inf_{t\in R}|x(t)| > 0$. If (6.1) has a solution bounded on $[t_o,\infty)$ then there is an almost periodic solution φ of (6.1) such that $\mathrm{mod}(\varphi) \subset \mathrm{mod}(A,f)$.

Proof: We first apply Theorem 6.2 to get a solution bounded on R and hence a bounded solution of all equations in the hull of (6.1). According to Theorem 5.1 each equation in the hull of (6.1) has a solution with minimum norm. This minimizing solution is unique, in fact, the remainder of the proof is now exactly the same as that of Theorem 5.8.

Corollary 6.4. If every bounded solution of an equation in the homogeneous hull is almost periodic, then all bounded solutions of (6.1) are almost periodic.

Proof: According to Theorem 5.7 the hypothesis of the Corollary imply the hypothesis of Theorem 6.3 so we get an almost periodic solution φ. But if ψ is another bounded solution, then $\psi = \varphi + \chi$ where χ is a solution of (6.2) so is almost periodic.

Note that the statement of Corollary 6.4 is in a form that includes the constant coefficient case. A natural question arises about the necessity of the hull hypothesis of either Theorem 6.3 or Corollary 6.4.

Question A. Is Theorem 6.3 true if only (6.2) is required to have $\inf |x(t)| > 0$ for nontrivial solutions?

Question B. Is Corollary 6.4 true with the hypothesis applying only to $x' = A(t)x$?

It is the author's conjecture that the answers are no and yes respectively.

There is a special case in which Theorem 6.3 is true without the hull hypothesis.

Corollary 6.5. If $A(t)$ is periodic, then any bounded solution of (6.1) is almost periodic.

Proof: According to Floquet Theory, there is a periodic matrix P so that P, P', and P^{-1} are bounded and the transformation $y = Px$ transforms (6.1) into

$$y' = Ry + Pf \tag{6.3}$$

and (6.2) into

$$y' = Ry \tag{6.4}$$

where R is a constant matrix. Now boundedness conditions on x carry over to the same ones for (6.3) and (6.4). So all bounded solutions of (6.4) are almost periodic and the hypotheses of Corollary 6.4 are satisfied by the system (6.3). This transforms back to the original equation since $x = P^{-1}y$ is almost periodic if y is almost periodic and P^{-1} is periodic.

The Floquet Theory for the almost periodic case, whatever that is supposed to be, will be discussed later.

5. Scalar equations. A little more can be said about scalar equations. This also applies to triangular vector systems. For if $A(t)$ is triangular, then one can examine a sequence of scalar equations, where the diagonal elements of A play the role of the coefficient A in the scalar version of (6.1). This reduction is outlined in the proof of Theorem 5.11.

Theorem 6.6. If $\text{Re } a(f,0) \neq 0$, then there is an almost periodic solution to the scalar equation $x' = f(t)x + g(t)$ if f and g are almost periodic. Furthermore, the solution has module contained in that of f and g.

Proof: Note that the homogeneous equation $x' = f(t)x$ has for its solutions $C \exp \int_0^t f(s)ds$. Since $\text{Re } a(f,0) \neq 0$, we have that

$\int_0^t f(s)ds$ is unbounded. If $a(f,0) > 0$, then $\int_0^T f(s)ds \to +\infty$ as $T \to \infty$ so all non-trivial solutions of $x' = f(t)x$ are unbounded. If $a(f,0) < 0$, then $\int_{-a}^{-a+T} f(s)ds \to -\infty$ as $T \to \infty$ uniformly in a so $\int_{-T}^{0} f(s)ds \to -\infty$, i.e. the solutions are unbounded as $t \to -\infty$. Now the condition that $a(f,0) \neq 0$ is shared by all equations in the hull, since the mean value is constant on the hull. Thus every equation in the hull of $x' = f(t)x + g(t)$ has a unique bounded solution, if it has one. To see that there is a bounded solution, assume for the sake of argument, that $f(t) = \lambda(t) + i\mu(t)$ where $\lambda_0 = M(\lambda) > 0$. The proposed bounded solution is

$$x(t) = \int_{-\infty}^{t} e^{-\int_s^t \lambda(u)du} e^{-i\int_s^t \mu(u)du} g(s)ds.$$

Note that

$$|x(t)| \leq \|g\| \int_{-\infty}^{t} e^{-\int_s^t \lambda(u)du} ds$$

$$= \|g\| \int_0^{\infty} e^{-\int_{t-v}^{t} \lambda(u)du} dv. \text{ But for}$$

$v \geq T_0$, $\left|\frac{1}{v}\int_{t-v}^{t} \lambda(u)du - \lambda_0\right| < \frac{\lambda_0}{2}$ by the existence of the mean value. Thus $-\int_{t-v}^{t} \lambda(u)du \leq -\frac{\lambda_0}{2}v$ if $v \geq T_0$. Consequently,

$$|x(t)| \le \|g\| \left[\int_0^{T_0} e^{-\int_{t-v}^{t} \lambda(u)\,du}\,dv + \int_{T_0}^{\infty} e^{-\frac{\lambda_0 v}{2}}\,dv \right]$$

$$\le \|g\| \left[T_0 e^{\|\lambda\| T_0} + \frac{2}{\lambda_0} e^{-\frac{\lambda_0 T_0}{2}} \right] < \infty .$$

A similar estimate justifies differentiation of the integral to show that x is in fact a solution. Now we can apply Theorem 6.3. To get module containment note that by uniqueness of the bounded solution, if α' is such that $T_{\alpha'} f = f$ and $T_{\alpha'} g = g$, then for some $\alpha \subset \alpha'$ $T_\alpha x = y$ exists uniformly. But y is a solution of the original equation so $y \equiv x$ and we get module containment by Theorem 4.5.

The above theorem handles the case of real part of the mean value non-zero, the case of zero real part of mean value is not so simple. For $g(t) \equiv 0$, we have a definitive statement. We shall prove that $f - a(f,0)$ must have bounded integral and that $a(f,0)$ must be pure imaginary. We begin by proving a Lemma.

<u>Lemma 6.7</u>. If $f(t) = \exp i\,F(t)$ is almost periodic for a real and continuous F, then there exist a real number c and an almost periodic G such that $F(t) = ct + G(t)$ with $\mathrm{mod}(G) \subset \mathrm{mod}(f)$.

<u>Proof</u>: If $0 < \epsilon < 1$ and $|f(x+\tau) - f(x)| < \epsilon$, then it follows that for some integral n $|F(x+\tau) - F(x) - 2n\pi| \le \epsilon\,\pi/2$. Since F is continuous, n is constant on connected sets where

$|f(x+\tau) - f(x)| < \epsilon$.

In particular, we get the uniform continuity of F. For if $0 < \epsilon < 1$, we have $|f(x+\tau) - f(x)| < \epsilon$ for all x provided $|\tau| \leq \delta$. If $\tau = 0$ then we have $|2n\pi| < \epsilon\pi/2 < \pi$, so $n = 0$ and $|F(x+\tau) - F(x)| < \epsilon\pi/2$ for $|\tau| \leq \delta$.

To construct c, note that it should in some sense be the average rate of growth of F. So consider $q(t,s) = \dfrac{F(t+s) - F(t)}{s}$ for $s > 0$. If $0 < \epsilon < 1$ and $\tau \in T(f,\epsilon)$ and $|\tau| \geq 1$, then $|f(x+\tau) - f(x)| < \epsilon$ for all x so that $|F(x+\tau) - F(x) - 2n\pi| < \epsilon\pi/2$ and n does not depend on x. If $s > \tau$, then write $s = m\tau + k$ with integral m and $0 \leq k < \tau$. Now $F(t+s) - F(t) = F(t+s) - F(t+m\tau)$ $+ \sum_{\upsilon=1}^{m} [F(t+\upsilon\tau) - F(t+(\upsilon-1)\tau)]$. From this, $|F(t+s) - F(t) - nm2\pi| \leq K + m\epsilon\pi/2$ where $K = \sup_{|x-y|\leq\tau} |F(x) - F(y)| < \infty$. Thus

$$\left|q(t,s) - \frac{nm2\pi}{m\tau + k}\right| \leq \frac{K}{s} + \frac{m\epsilon\pi/2}{s} < \frac{K}{s} + \frac{\epsilon\pi}{2}$$

which we can write as

$$\left|q(t,s) - \frac{2\pi n}{\tau + k/m}\right| < \frac{K}{s} + \frac{\epsilon\pi}{2} .$$

This shows that given $0 < \epsilon < 1$, there is an S such that for $s > S$

$$\sup_t q(t,s) - \inf_t q(t,s) < 3\epsilon\pi/2.$$

Thus $\lim_{s\to\infty} q(t,s) = c$ exists uniformly in t and is a constant.

Let $g(t) = F(t) - ct$, then $e^{ig(t)} = e^{-ict}e^{iF(t)}$ is almost periodic and the c corresponding to g is zero. Thus without loss of generality, we may assume that $c = 0$ for the remainder of the proof.

If $\tau \in T(f,\epsilon)$, for $0 < \epsilon < 1$ and $\tau \neq 0$, let $n(\tau)$ be such that $|F(x+\tau) - F(x) - 2n(\tau)\pi| < \epsilon\pi/2$ and $n(\tau)$ does not depend on x. If $n(\tau) \neq 0$, then either $F(x+\tau) - F(x) > \pi$ for all x or $F(x+\tau) - F(x) < -\pi$ for all x. In either case, an easy induction gives $|F(x+n|\tau|) - F(x)| > n\pi$ for all x. Thus

$$0 = |c| = \lim_{n\to\infty} \frac{F(t+n|\tau|) - F(t)}{n|\tau|} \geq \frac{\pi}{|\tau|} > 0.$$

We conclude that $n(\tau) = 0$, and $T(F,\epsilon\pi/2) \supset T(f,\epsilon)$.

<u>Theorem 6.8</u>. Let $f \in AP(C)$. Then $\exp \int_a^x f(s)ds \in AP(C)$ if and only if $f(t) = ic + g(t)$ where c is real and $\int_a^x g(t)$ is bounded (and $g \in AP(C)$).

<u>Proof</u>: If f is of the given form then $\exp \int_a^x f$ certainly is almost periodic. It is the converse that concerns us here. First, if f is real then $\exp \int_a^x f(s)ds$ is an almost periodic solution to $y' = fy$ so $\inf_t |y(t)| > 0$ and $\sup_t |y(t)| < \infty$. Thus $\int_a^x f(s)ds$ must be bounded and hence almost periodic. Then $c = 0$ is the required constant. If f is complex, write

$\int_a^x f = F_1 + iF_2$ and $f = f_1 + if_2$ with $F_i' = f_i$. Then $h(t) = \exp \int_a^t f(s)ds = e^{F_1} e^{if_2}$. If $h(t)$ is almost periodic, so is $|h(t)|$. Thus e^{F_1} is almost periodic and by the first part F_1 is bounded. Then $\frac{h(t)}{|h(t)|} = e^{iF_2}$ is almost periodic since $\inf_t |h(t)| > 0$. By the preceding lemma, $F_2 = ct + G_2$ with $G_2 \in AP(R)$. Now $f_2 = c + G_2'$ so G_2' is almost periodic. Thus $g = f_1 + i(f_2 - c)$ is an almost periodic function with the almost periodic integral $F_1 + iG_2$. Then $f = ic + g$ has the required form.

<u>Corollary 6.9</u>. Let $\operatorname{Re} a(f,0) = 0$. The equation $x' = f(t)x$ has all its solutions almost periodic if and only if $f(t) - a(f,0)$ has bounded integral.

<u>Proof</u>: If all solutions are in $AP(C)$, then $\exp \int_0^x f(s)ds$ is almost periodic, so $f(t) = ic + g(t)$ with $\int_0^x g$ bounded. This implies that $a(g,0) = 0$ so $a(f,0) = ic$ thus $g = f - a(f,0)$ has bounded integral. Conversely, if $f - a(f,0)$ has bounded integral then $f = ic + g(t)$ where g has bounded integral and Theorem 6.8 applies. Here $ic = M(f)$.

6. <u>Linear homogeneous systems</u>. Our previous results are concerned with equations which have a bounded solution. At the other end of the spectrum is the situation when all solutions are bounded. For the homogeneous case

$$x' = A(t)x \qquad (6.5)$$

this is a condition on the fundamental solution.

Theorem 6.10. Suppose $\int_0^t \operatorname{trace} A(s)\,ds$ is bounded, and suppose there is a fundamental set of solutions $x_1,\ldots,x_n$ which are bounded on R. Let x_1 satisfy the condition that if φ is a solution such that $\operatorname{Range}(\varphi) \subset \overline{\operatorname{Range}(x_1)}$, then $\varphi \equiv x_1$. Then x_1 is almost periodic, and $\operatorname{mod}(x_1) \subset \operatorname{mod}(A)$.

Proof: Note that trace $A(s)$ is almost periodic so $\int_0^t \operatorname{trace} A(s)\,ds$ is almost periodic and thus the Wronskian $W(x_1,\ldots,x_n)(t) = W(x_1,\ldots,x_n)(0)\exp \int_0^t \operatorname{trace} A(s)\,ds$ is also almost periodic. Furthermore, $|W(x_1,\ldots,x_n)(t)| \geq k > 0$ for some constant k. The plan of the proof is to show there is a fundamental set for every equation in the hull of (6.5) that satisfies the hypothesis of the theorem where x_1 is replaced by $T_\alpha x_1$. So suppose $T_\alpha A = B$ uniformly and $T_\alpha x_i = y_i$ exist uniformly on compact subsets of R. Then $\operatorname{Range}(y_i) \subset \overline{\operatorname{Range}(x_i)}$ so we have a bounded set of solutions of $x' = B(t)x$. Furthermore, $W(y_1,\ldots,y_n)(t) = T_\alpha W(x_1,\ldots,x_n)$ must also be an almost periodic function. Hence $|W(y_1,\ldots,y_n)(t)| \geq k > 0$ so that $y_1,\ldots,y_n$ form a fundamental set. Now suppose ψ is a solution of $x' = B(t)x$ such that $\operatorname{Range}(\psi) \subset \overline{\operatorname{Range}(y_1)}$. Then $\psi = c_1y_1 + \ldots + c_ny_n$. Let $\alpha' \subset \alpha$ such that $z_i = T_{-\alpha'}y_i$ exist uniformly on compact subsets of R. Then $T_{-\alpha'}B = A$ because $T_{-\alpha}B = A$ and by the above argument $z_1,\ldots,z_n$ form a fundamental set of solutions of $x' = A(t)x$. Now

$\mathrm{Range}(z_1) \subset \overline{\mathrm{Range}(y_1)} \subset \overline{\mathrm{Range}(x_1)}$ so $z_1 \equiv x_1$. It follows that $T_{-\alpha'}\psi = \varphi$ exists and $\mathrm{Range}(\varphi) \subset \overline{\mathrm{Range}(\psi)} \subset \overline{\mathrm{Range}(x_1)}$ so $\varphi \equiv x_1$. But $x_1 = \varphi = T_{-\alpha'}\psi = c_1 T_{-\alpha'} y_1 + \ldots + c_n T_{-\alpha'} y_n$

$$= c_1 x_1 + c_2 z_2 + \ldots + c_n z_n.$$

By independence it must be that $c_1 = 1$ and $c_2 = \ldots = c_n = 0$. That is $\psi = y_1$. It follows also from the above, that $\overline{\mathrm{Range}(y_1)} = \overline{\mathrm{Range}(x_1)}$. In particular, if B is in the hull of A such that $T_\alpha A = B$, there might be two subsequences $\alpha_1 \subset \alpha$ and $\alpha_2 \subset \alpha$ such that $T_{\alpha_1} x_1$ and $T_{\alpha_2} x_1$ exist uniformly on compact subsets. Now from the range condition $T_{\alpha_1} x_1$ and $T_{\alpha_2} x_1$ have the same closure of range so are identical.

We can now conclude the argument for the almost periodicity in a familiar way. If α' and β' are given, we extract common subsequences α and β so that $T_{\alpha+\beta} A = T_\alpha T_\beta A$; $T_{\alpha+\beta} x_1$, $T_\alpha T_\beta x_1$ all exist uniformly on compact sets. Then the latter two are solutions of the same equation with same Range so are identical. Thus x_1 is almost periodic. It is also clear that if $T_{\alpha'} A = A$ and $\alpha \subset \alpha'$ such that $T_\alpha x_1 = y_1$ then $\mathrm{Range}(y_1) \subset \overline{\mathrm{Range}(x_1)}$ so $y_1 = x_1$ and thus $\mathrm{mod}(x_1) \subset \mathrm{mod}(A)$.

Note that if more than one of the solutions satisfy the hypothesis on x_1, then these solutions are also almost periodic. In particular, if all the x_i have the range condition, then the fundamental solution is almost periodic. Note also that the condition that $\int_0^t \mathrm{trace}\, A(s)\,ds$ be bounded is equivalent to $|W(x_1,\ldots,x_n)(t)| \geq k > 0$.

A question about almost periodic fundamental solutions that is important in later considerations is the following. When is the inverse also almost periodic? We can establish a result that shows that the hypothesis on the Wronskian (or equivalently the trace A) found in the previous two results is a natural one.

<u>Theorem 6.11</u>. Let A and B be almost periodic matrices, and suppose that the fundamental matrix solution $X(t)$ to

$$X' = AX + XB \qquad X(0) = I \tag{6.6}$$

is almost periodic. Then X^{-1} is also almost periodic.

<u>Proof</u>: The general matrix case can be reduced to the scalar case in the following way. Since $X^{-1}(t) = (\det X)^{-1} (\text{adjoint } X)^T$ it is sufficient to prove that $(\det X)^{-1}$ is almost periodic. Because each component of adjoint X is a polynomial in the components of X, adjoint X is almost periodic, and for the same reason $\det(X)$ is almost periodic. Now if $Y' = AY$ $Y(0) = I$, and $Z' = ZB$ $Z(0) = I$ then it is easy to check that $X = YZ$ and $\det(X) = \det(Y)\det(Z)$. Now $\det(Y)' = (\text{trace } A)\det(Y)$ and $(\det Z)' = (\text{trace } B)\det(Z)$. Consequently $x = \det(X)$ satisfies

$$x' = (\text{trace } A + \text{trace } B)x = \text{trace}(A+B)x \qquad x(0) = 1.$$

According to Theorem 5.7, $\inf|x(t)| > 0$ so $|x(t)|^{-1}$ is almost periodic as required.

Note that the almost periodicity of x in the above proof

requires that $\exp\int_0^t \text{trace}(A+B)(s)ds$ must be almost periodic. According to Theorem 6.8 $\text{trace}(A+B) = ic + g(t)$ where g is almost periodic with bounded integral. This is a necessary condition for the existence of a fundamental almost periodic solution.

A related question might be the question of when a matrix almost periodic function is in fact non-singular. This is a meaningful question when the solution is constructed in a way that yields the fourier series instead of its values at some point.

Theorem 6.12. If X is a matrix almost periodic solution to the almost periodic system (6.5), then X is a non-singular solution if one of its fourier coefficients is non-singular.

Proof: Suppose $X(0)$ is singular and let Y be the fundamental solution such that $Y(0) = I$. Then $X(t) = Y(t)X(0)$. Thus X has its values in the Banach algebra of matrices $D = \{CX(0) : C \text{ is an } n \times n \text{ matrix}\}$. It follows that $e^{i\lambda t}X(t) \in D$ for all λ and t and hence $\sum_{i=1}^{N} e^{i\lambda t_i}X(t_i)\Delta t_i \in D$ and these Riemann sums converge to $\int_0^t e^{i\lambda s}X(s)ds$. Since D is closed, this integral is in D and thus $M\{e^{i\lambda t}X(t)\} \in D$. But D consists of singular matrices, so all fourier coefficients of X are singular. The theorem follows by contraposition.

The converse of this theorem is not true since a diagonal matrix can be non-singular with all the fourier coefficients singular. $\begin{pmatrix} e^{it} & 0 \\ 0 & e^{2it} \end{pmatrix}$ is an example. However it would be of interest to know if Theorem 6.12 can be extended to the system (6.6).

Question C. Can Theorem 6.13 be extended to the system (6.6)?

7. Periodic Homogeneous Systems. Slightly more can be said for some periodic systems of the form (6.5). Of course the classical Floquet Theory gives some information. Here we are considering some special cases. Assume that A is skew symmetric and real. Then the fundamental solution to (6.5) is symmetric. To see this, note that $(XX^T)' = X'X^T + X(X')^T = AXX^T - XX^TA^T$ so that XX^T is a solution to $U' = AU - UA$ and $U(0) = I$. But I is also a solution of this system so that $XX^T = I$. It is easy to see that if $XX^T = I$ for all t, then A is skew symmetric.

Theorem 6.13. If A is periodic and skew symmetric, then the fundamental solution of (6.5) is almost periodic.

Proof: All vector solutions of (6.5) are bounded. Indeed if x is a solution then $x(t) = X(t)\xi$ so that $|x(t)|^2 = \xi^T X^T(t) X(t) \xi = |\xi|^2$. By Corollary 6.5, $x(t)$ is almost periodic.

Question D. Is Theorem 6.13 true for almost periodic A?

Note that trace $A = 0$ so that the necessary condition for X to be almost periodic is satisfied. In fact, if A is complex and skew-hermitian, then X is hermitian and trace $A =$

pure imaginary and Theorem 6.13 is true in this case also.

By Floquet Theory we may look at Theorem 6.13 in a slightly different way. That theory says that $X(t) = P(t)e^{Rt}$ where P is periodic with R a constant. Thus $e^{Rt} = P^{-1}(t)X(t)$ is almost periodic, and its inverse e^{-Rt} is also almost periodic.

It would be of interest to look at the general case of periodic A by factoring the solution in a different way. Every matrix $A = K + S$ where K is skew-symmetric and S is symmetric. Let $y' = Ky$, $y(0) = I$ and $z' = \frac{1}{2}(y^T Sy)z$, $z(0) = I$. Then it is a simple exercise to show that (yz) is a solution to $X' = AX$ $X(0) = I$. $(yy^T = I)$ so yz is the fundamental solution of (6.5). We know that y is almost periodic, what about z?

<u>Theorem 6.14</u>. Let z be defined as above, then $z(t) = F(t)e^{Dt}$ where F is almost periodic and D is a constant.

<u>Proof</u>: This is easy to see, since $X(t) = P(t)e^{Rt}$ and $X = Yz$. Then $z = Y^{-1}P(t)e^{Rt}$ and $Y^{-1}(t)P(t)$ is almost periodic.

Actually, Theorem 6.14 does not help to represent solutions of the general equation (6.5) since Floquet Theory does that. However Theorem 6.13 and 6.14 recover the solution of (6.5) from two special cases. If these two were true for certain almost periodic systems, then one could derive a Floquet Theory for almost periodic systems of that type.

<u>Question E</u>. If A is almost periodic and symmetric, is the fundamental solution of $X' = AX$ of the form $F(t)e^{Dt}$ for F almost periodic and D constant? Or is there a subclass for

which this is true, or is there a subclass for which this is true for F merely bounded?

8. Homogeneous Systems with Bounded Fundamental Solutions. We saw in the previous section that for skew-symmetric A all solutions are bounded. One might ask for other conditions under which all solutions are bounded. We can take the constant coefficient case as an example. If A is a constant, then all solutions are bounded if and only if all the roots of A are pure imaginary with simple elementary divisors. If J is the Jordan canomial form, then $0 = J + J^* = PAP^{-1} + (P^*)^{-1}A^*P^*$ or $(P^*P)A + A^*(P^*P) = 0$. Now $Q = P^*P$ is a positive definite matrix such that $QA + A^*Q = 0$. This condition can be generalized to the time dependent case by allowing Q to depend on time, and $0 = Q'$ in the constant case, is replaced in that way.

Theorem 6.15. If A is continuous then the following are equivalent.

(1) there is a $Q(t)$ and $0 < r < \delta$ such that $QA + A^*Q = - Q'$ and $r|\xi|^2 \leq \xi^*Q\xi \leq \delta|\xi|^2$ for all $\xi \in C^n$.

(2) there is an α such that $|x(t)|^2 \leq \alpha|x(t_o)|^2$ for all vector solutions of (6.5).

Proof: To prove that (1) implies (2), let x be any vector solution. It is a straight forward computation to show that $(x^*Qx)' = 0$ so that x^*Qx is a constant. Then $r|x(t)|^2 \leq x^*Qx(t) = x^*Qx(t_o) \leq \delta|x(t_o)|^2$ so $\alpha = \delta/r$ works.

Conversely, let $X(t,\tau)$ be the solution of (6.5) such that $X(T,\tau) = I$. Recall that as a function of τ, $X(t,\tau)$ is a solution of the adjoint equation $X' = -XA$. The proposed matrix is $Q(t) = X^*(\tau,t)X(\tau,t)$ with some fixed τ. A simple computation gives $Q' = -QA - A^*Q$ so we need only show it is uniformly definite. Note that $\xi^*Q\xi = |X(\tau,t)\xi|^2$. But $X(\tau,t)\xi$ as a function of τ is a solution of $X' = AX$ so

$$\xi^*Q\xi \le \alpha|X(t,t)\xi|^2 = \alpha|\xi|^2.$$

Also $|\xi|^2 = |X(t,\tau)X(\tau,t)\xi|^2 \le \alpha|X(\tau,t)\xi|^2 = \alpha\xi^*Q\xi$. Thus $r = \alpha^{-1}$ and $\delta = \alpha$.

To tie this in with the previous discussion, if $y = Px$ is a change of variables to $y' = By$ with B skew-symmetric, then $Q = P^*P$ is the Q of the theorem. On the other hand if there is a differentiable P such that $Q = P^*P$ then B will be skew-symmetric.

9. Contribution to Question E. There is a special situation in which one can prove a representation of solution to almost periodic systems which parallels that for the Floquet Theory representation for periodic systems.

Definition 6.16. The semi-module of an almost periodic function f is the smallest set which contains $\exp(f)$ and is closed under addition. Denote the semi-module by $S(f)$.

<u>Theorem 6.17</u>. Suppose A is a constant matrix and B an almost periodic matrix such that if λ_i are the eigenvalues of A, then $\mathrm{dist}(\lambda_k - \lambda_j, i\,S(B)) > 0$, all j and k. Then the fundamental solution $X(t)$ of $x' = (A + B(t))x$ is of the form $X(t) = [I + Q(t)]e^{At}$ where Q is almost periodic and $\exp(Q) \subset S(B)$.

<u>Proof</u>: Change variables by $X = Pe^{At}$ to get

$$P' = AP - PA + B(t)P. \tag{6.7}$$

We look at this type of equation. Let $C(t)$ be almost periodic with $\exp(C) \subset S(B)$, and consider

$$U' = AU - UA + C(t). \tag{6.8}$$

The linear map $U \to AU - UA$ has eigenvalues of the form $\lambda_i - \lambda_j$ so we may apply Theorem 5.11 to get a unique almost periodic solution U of (6.8) whose exponents are the same as C, and the solution U satisfies $\|U\| \leq p(D\rho^{-1})\|C\|$ where ρ is the distance of $\{\lambda_k - \lambda_j\}$ to $i\exp(C)$. We will use this repeatedly. It is easy to show that $S(B)$ is dense in R if it contains numbers of opposite signs and if B is not periodic. So assume B is not periodic and thus $\exp(B)$ are all one sign. In particular B must be complex. We solve (6.7) by iteration in the following way. Let $U_o = I$ and define U_n as the unique almost periodic solution of (6.8) with $C(t) = B(t)U_{n-1}$ and $\exp(U_n) = \exp(B(t)U_{n-1}(t)) \subset \exp(B) + \exp(U_{n-1})$. Thus $\exp(U_n) \subset S(B)$ for all n and this iteration is well defined.

Since $0 = \lambda_j - \lambda_j$ is a positive distance from $i\,S(B)$, if $\lambda \in S(B)$ then $|\lambda| \geq d$. Now any exponent of U_n, μ say, is of the form $m_1\lambda + m_2\sigma$ where $\lambda \in \exp(B)$ and $\sigma \in \exp(U_{n-1})$, m_1 and m_2 are positive integers. To see this, consider the formal product of the fourier series. In particular we have that $|\mu| \geq nd$. Eventually we have that $d_n = \operatorname{dist}(i\exp U_n, \lambda_k - \lambda_j)$ is as large as one likes. We have the estimate for sufficiently large n

$$\|U_n\| \leq M\,p(D\,d_n^{-1})\,\|B\|\;\|U_{n-1}\|$$

We may choose N so large that $M\,p(D\,d_n^{-1})\,\|B\| < \frac{K}{n}$ for $n \geq N$, since p is a polynomial with no constant term and d_n grows like n. From $\|U_n\| \leq \frac{K}{n}\|U_{n-1}\|$ we get that $\sum \|U_n\|$ is dominated by $\sum \frac{K^n}{n!}$. Thus the series

$$P(t) = \sum_{n=o}^{\infty} U_n(t) = I + Q(t)$$

converges uniformly on R and is almost periodic with $\exp(Q) \subset S(B)$. It is easy to verify that P is a solution of (6.7).

In the same way, we can find an almost periodic solution to $\tilde{P} = I + \tilde{Q}$

$$P' = AP - PA - P\,B(t) \qquad (6.9)$$

with $\exp\tilde{Q} \subset S(B)$. Then $\tilde{P}P$ is a solution of

$$P' = AP - PA \qquad (6.10)$$

Another solution is I. Therefore $\tilde{P}P - I = \tilde{Q} + Q + \tilde{Q}Q$ is a solution of (6.10) with exponents in $S(B)$. But zero is a solution of (6.10) with exponents in $S(B)$ so by uniqueness, $\tilde{P}P - I = 0$ and P is invertible so is a fundamental solution.

Note that a version of Theorem 6.12 would be helpful here.

10. Notes. The results of article 4 can be found in Favard's book [172] and in [179]. The exception is Corollary 6.4 which is in Bochner, A new approach to almost periodicity, Proc. Nat. Acad. Sci. USA, 46(1960), 1233-1236. This latter paper has revolutionized the indirect approaches that we use in this book. The other influential paper is that of Amerio [6] which we cite later.

The reader might refer to Example 11.19 as shedding some light on Questions A and B.

The result of Theorem 6.6 is found in Cameron, Quadratures involving trigonometric sums, J. Math. Phys. 19(1940), 161-166. The extension to triangular systems is noted in Massera, Un criterio de existencia de soluciones casiperiodicas de ciertos sistemas de ecuaciones diferencialas casi-periodicas, Publ. Inst. Mat. y Estad. Fac. Ing. y Agrimensura, Montevideo, 3(1958), 89-102.

One of the more interesting early investigations into almost periodic solutions has to do with the problem of mean motions. Theorem 6.7 is an example where the decomposition $F(t) = ct + G(t)$ shows that F oscillates about the linear motion ct. This result can be found in Bohr's works. A discussion of the entire problem of mean motions can be found in

Jessen, Some aspects of the theory of Almost Periodic Functions, Proc. Int. Congress Amsterdam 2(1954), 1-10.

The extension of 6.7 to Theorem 6.8 can be found in Bochner, Remark on the integration of almost periodic functions, J. London Math. Soc. 8(1933), 250-254. The result of Theorem 6.10 is also due to Bochner, and can be found in, Homogeneous systems of differential equations with almost periodic coefficients, J. London Math. Soc. 8(1933), 283-288. More interesting results on homogeneous systems can be found in the papers of Bochner and Cameron. These references may be found in Chapter 15.

Theorem 6.11 can be found in Fink, Almost Periodicity of the inverse of a fundamental matrix, Proc. Amer. Math. Soc. 27(1971), 527-529, and the related result in 6.12 in Frederickson, Generalized series solution of an almost periodic differential equation, J. Differential Equations 2(1966), 243-264. This paper is an attempt to use summation methods to construct almost periodic solutions.

The results of Theorems 6.13, 6.14 and 6.15 can be found in the papers (respectively) van Vleck, A note on the relation between periodic and orthogonal fundamental solutions of linear systems, Amer. Math. Monthly 77(1964), 774-776; Burton, Linear differential equations with periodic coefficients, Proc. Amer. Math. Soc. 17(1966), 327-329; Schaeffer, Boundedness of solutions to linear differential equations, Bull. Amer. Math. Soc. 74(1968), 508-511.

Theorem 6.17 is found in Coppel, Almost periodic properties

of ordinary differential equations, Ann. Mat. Pura Appl. 76(1967), 27-50. Other results along these lines may be found in the papers of Golomb, see Chapter 15.

There are also results about integration of almost equations in Banach spaces. These have been studied by Amerio and his students, see for example, [24].

Chapter 7

Exponential Dichotomy and Kinematic Similarity

1. Introduction. A glance at article 5.11 shows that if A is constant and there is a bounded solution of $x' = Ax+f$ for every bounded f, then there is a projection P so that $|e^{At}P| \leq K_1 e^{-\sigma_1 t}$ for $t \geq 0$ and $|e^{At}(I-P)| \leq K_2 e^{\sigma_2 t}$ for $t \leq 0$ where σ_1 and σ_2 are positive. Such estimates are called exponential dichotomies. If A is non-constant and almost periodic, then in order to get almost periodic solutions to $x' = Ax+f$ one needs estimates for the fundamental solutions of $x' = Ax$. Such estimates are at the opposite extreme from having all solutions bounded. Here only 0 is a bounded solution.

2. Exponential Dichotomy.

Definition 7.1. The equation $x' = A(t)x$ satisfies an exponential dichotomy if there exists a projection P and positive constants $\sigma_1, \sigma_2, K_1, K_2$ so that

$$\begin{aligned} |X(t)PX^{-1}(s)| &\leq K_1 e^{-\sigma_1(t-s)} \qquad t \geq s \\ |X(t)(I-P)X^{-1}(s)| &\leq K_2 e^{-\sigma_2(t-s)} \qquad t \leq s \end{aligned} \tag{7.1}$$

for a fundamental solution X.

It is of interest to look at exponential dichotomy in light of Floquet Theory for periodic systems. If A is periodic, then $X(t) = Q(t)e^{Bt}$ where Q is periodic and B constant. It is thus clear that one has an exponential dichotomy if and only if the matrix B has all its eigenvalues off the imaginary axis. This is the same property that the equation $x' = Bx$ has. The two equations are related in the following way. Take the equation $x' = A(t)x$ and change variables by $x = Q(t)y$ and require that y satisfy $y' = By$. Then Q must satisfy the equation

$$Q' = AQ - QB \tag{7.2}$$

As in Floquet Theory we want boundedness to carry over under this change of variable. We require that both Q and Q^{-1} are bounded. Note that if Q is a constant then (7.2) just gives B being similar to A. The generalization to non-constant A is called kinematic similarity.

Definition 7.2. The equations $x' = A(t)x$ and $x' = B(t)x$ are said to be kinematically similar if there exists a matrix Q, such that Q and Q^{-1} are bounded and satisfy (7.2).

To review, two equations that are kinematically similar have the same boundedness and stability properties.

Floquet Theory, in this terminology, then is that every periodic system is kinematically similar to a constant coefficient equation via a periodic matrix Q. It would be nice if the same statement for almost periodic equations were true. It is not, but we postpone the example. In the Russian literature, kinematic similarity to a constant coefficient system is called reducible. In any case one would like $B(t)$ to be nicer in some sense than A.

Our first formal observation is that two kinematically similar equations satisfy common exponential dichotomies.

Theorem 7.3. If $x' = Ax$ satisfies an exponential dichotomy (7.1) and is kinematically similar to $x' = Bx$, then this latter equation satisfies an exponential dichotomy with the same constants σ_i and same projection P.

Proof: Let $Y(t) = Q^{-1}X$ and apply 7.1. Only the K's change.

3. Exponential Dichotomy of almost periodic systems. We are basically going to show that exponential dichotomy of an almost periodic equation carries over to the equations in the hull. We first need two lemmas.

Lemma 7.4. Let P be a projection and X a differentiable invertible matrix such that XPX^{-1} is bounded on R. There exists a differentiable S such that $XPX^{-1} = SPS^{-1}$ for all $t \in R$ and S and S^{-1} are bounded on R. In fact, there is an S of the form $S = XQ^{-1}$ where Q commutes with P.

Proof: One can reduce the general case to the case where P is an orthogonal projection by a similarity transformation. We will use $|A| = (\operatorname{trace} A^*A)^{1/2}$ for a norm in this proof. The matrix $Q^2 = PX^*XP + (I-P)X^*X(I-P)$ is a positive definite hermitian matrix so has a square root which is positive definite and hermitian. It is easy to verify that Q^2 commutes with P and so does Q, being a polynomial in Q^2. Now from the definition of Q we have $I = (Q^{-1}PX^*)(XPQ^{-1}) + (Q^{-1}(I-P)X^*)(X(I-P)Q^{-1})$ so that $n = |XPQ^{-1}|^2 + |X(I-P)Q^{-1}|^2$ where n is the size of the matrices. Thus $|XQ^{-1}| = |XPQ^{-1} + X(I-P)Q^{-1}| \leq |XPQ^{-1}| + |X(I-P)Q^{-1}| \leq 2\sqrt{n}$. Also $|QX^{-1}|^2 = |XPX^{-1}|^2 + |X(I-P)X^{-1}|^2 \leq 2|XPX^{-1}|^2 + n^2 \leq K$ by the hypothesis. Now $S = XQ^{-1}$ is bounded with bounded inverse. Since Q^2 is differentiable so is Q.

<u>Lemma 7.5</u>. If $x' = Ax$ satisfies an exponential dichotomy (7.1) and X is the fundamental solution, then XC, C non-singular, satisfies the exponential dichotomy with the same projection P if and only if $CP = PC$.

<u>Proof</u>: If ξ is a vector then $|X(t)(I-P)X^{-1}(s)\xi| \leq |\xi| K_2 e^{-\sigma_2(t-s)}$, $t \leq s$. Take $\xi = X(s)\eta$ where η is an arbitrary vector to get $|X(s)\eta| \geq K_2^{-1} e^{\sigma_2(t-s)} |X(t)(I-P)\eta|$ if $t \leq s$. If $(I-P)\eta \neq 0$, then $|X(s)\eta| \to +\infty$ as $s \to \infty$. But if $P\eta = \eta$ then $|X(t)\eta| = |X(t)P\eta| = |X(t)PX^{-1}(s)\xi| \leq K_1 e^{-\sigma_2(t-s)} |\xi|$ if $t \geq s$ and so $|X(t)\eta| \to 0$ as $t \to \infty$.

A similar proof shows that $|X(t)\eta| \to \infty$ as $t \to -\infty$ if $P\eta \neq 0$ and $|X(t)\eta| \to 0$ as $t \to +\infty$ if $P\eta = \eta$. Now if $X(t)C$ satisfies the exponential dichotomy with same P then $X(t)CP$ is bounded as $t \to \infty$. By above, $P(CP) = CP$. In same way $X(t)C(I-P)$ is bounded as $t \to -\infty$ so by above $PC(I-P) = 0$. Thus $PC = PCP = CP$.

<u>Theorem 7.6</u>. Suppose the system $x' = A(t)x$ with A almost periodic,

satisfies an exponential dichotomy. If $B \in H(A)$ then $x' = Bx$ satisfies an exponential dichotomy with same projection and same constants.

Proof: We invoke Lemma 7.4 where X is the fundamental solution satisfying (7.1). Let Q and S be given by that Lemma. Let $T_\alpha A = B$ uniformly and let $X_n(t) = X(t+\alpha_n)Q^{-1}(\alpha_n)$. This is a fundamental solution to $x' = A(t+\alpha_n)x$ and satisfies (7.1) with same projection P and same constants. This is true since Q^{-1} commutes with P. We may take subsequences so that $X_n(0)$ and $X_n^{-1}(0)$ converge since they are $S(\alpha_n)$ and $S^{-1}(\alpha_n)$ and these are bounded. Without changing notation we may assume $X_n(0) \to Y_0$ and hence $X_n^{-1}(0) \to Z_0$ where $Z_0 = Y_0^{-1}$. But now for a suitable subsequence $X_n(t)$ converges to a solution of $x' = Bx$ uniformly on compact intervals. Call this solution Y. Then $Y(0) = Y_0$ so is non-singular and clearly Y satisfies (7.1) since X_n does for all n.

If the homogeneous equation satisfies an exponential dichotomy then there is only one bounded solution, namely 0. By Theorem 7.6 this is also true for all equations in the hull. Thus the bounded solutions of the non-homogeneous equation should be almost periodic.

Theorem 7.7. Suppose the homogeneous equation $x' = A(t)x$ satisfies an exponential dichotomy (7.1) and f is an almost periodic function. Then there is a unique almost periodic solution φ to $x' = A(t)x + f(t)$ and $\mathrm{mod}(\varphi) \subset \mathrm{mod}(A,f)$. Furthermore,

$\|\varphi\| \leq \left(\frac{K_1}{\sigma_1} + \frac{K_2}{\sigma_2}\right)\|f\|$ where K_i and σ_i are the constants in (7.1).

Proof: We note that

$$\varphi(t) = \int_{-\infty}^{t} X(t)PX^{-1}(s)f(s)ds - \int_{t}^{\infty} X(t)(I-P)X(s)f(s)ds$$

is formally a solution and the differentiation is justified by computations similar to

$$\|\varphi\| \leq \left[\int_{-\infty}^{t} |X(t)PX^{-1}(s)|ds + \int_{t}^{\infty} |X(t)(I-P)X^{-1}(s)|ds\right]\|f\|$$

$$\leq \left[\int_{-\infty}^{t} K_1 e^{-\sigma_1(t-s)} ds + \int_{t}^{\infty} K_2 e^{\sigma_2(t-s)}\right]\|f\|$$

$$= \left(\frac{K_1}{\sigma_1} + \frac{K_2}{\sigma_2}\right)\|f\|.$$

Since the homogeneous equation has no non-trivial solutions, x is the unique bounded solution. Furthermore, every equation in the hull has a unique bounded solution. Thus if α' and β' are given, we find $\alpha \subset \alpha'$, $\beta \subset \beta'$, common subsequences so that $T_{\alpha+\beta}A = T_\alpha T_\beta A$, $T_{\alpha+\beta}f = T_\alpha T_\beta f$, $y = T_{\alpha+\beta}\varphi$, and $z = T_\alpha T_\beta \varphi$ exist uniformly on compact sets. But $y = z$ since both are bounded solutions of the same equation. So φ is almost periodic. If $T_\alpha A = A$ and $T_\alpha f = f$, then $T_\alpha \varphi = \varphi$ so $\mathrm{mod}(\varphi) \subset \mathrm{mod}(A,f)$.

4. More on Exponential Dichotomy. It is easy to see that exponential dichotomy is in fact equivalent to the eigenvalues of A being off the imaginary axis if A is constant. In fact, if all eigenvalues

have negative real parts, then solutions decay to 0 exponentially as $t \to \infty$. The situation for non-constant coefficients is quite different. Consider the following example.

Example 7.8. Let $A(t) = \begin{pmatrix} -1 + \frac{3}{2}\cos^2 t & 1 - \frac{3}{2}\cos t \sin t \\ -1 - \frac{3}{2}\sin t \cos t & -1 + \frac{3}{2}\sin^2 t \end{pmatrix}$.

It is easy to verify that for all t, the eigenvalues of $A(t)$ are the constants $\frac{1}{4}(-1 \pm \sqrt{7}\, i)$. Yet $(-\cos t, \sin t)e^{t/2}$ is a solution whose norm increases to $+\infty$ as $t \to +\infty$.

There are several known conditions that give exponential dichotomy however. One which we shall state without proof is given by Coppel [ ].

Theorem 7.9. Let $A(t)$ be differentiable and $|A(t)| \leq M$. If $A(t)$ has k eigenvalues whose real part $\leq -\alpha < 0$ and $n-k$ eigenvalues whose real part $\geq \beta > 0$, then for $0 < \epsilon < \min(\alpha,\beta)$ there is a constant δ, depending only on M, $\alpha+\beta$, and ϵ, such that if $|A'(t)| < \delta$ then $x' = Ax$ satisfies an exponential dichotomy (7.1) where $P = \begin{pmatrix} I_k & 0 \\ 0 & 0 \end{pmatrix}$ and K_i depend only on M, $\alpha+\beta$, and ϵ.

Apparently, the example 7.8 has $A'(t)$ too large. However, there are conditions that do not require smallness of A' but are explicit conditions on the matrix A. As an introduction, we noted that it is not sufficient to require that the eigenvalues of $A(t)$ are off the imaginary axis. There is a well known theorem of Geschorgin whose conclusion is that the eigenvalues are off the imaginary axis in certain cases. This is the following.

Geschorgin's theorem. Let $r_i = \sum_{j \neq i} |a_{ij}|$. Then the eigenvalues of A are found in the union of the discs in the complex plane centered at a_{ii} with radii r_i.

It follows that if $|\mathrm{Re}(a_{ii})| > r_i$, then all of the eigenvalues of A are off the imaginary axis. This fact, as we have seen above, is not sufficient for exponential dichotomy. However, the remarkable thing is that the condition that $|\mathrm{Re}(a_{ii})| > r_i$ is sufficient. Note that A^T has the same eigenvalues as A so that we also have an analogous condition where r_i is replaced by $\sum_{j \neq i} |a_{ji}|$.

Definition 7.10. If $A(t)$ is a matrix so that $|\mathrm{Re}\, a_{ii}(t)| \geq \sum_{j \neq i} |a_{ij}(t)| + \delta$ for some $\delta > 0$ and all i, then A is said to be row dominant. If $|\mathrm{Re}\, a_{ii}(t)| \geq \sum_{j \neq i} |a_{ji}(t)| + \delta$ is column dominant.

5. Row dominant systems. We will show by a sequence of lemmas that row dominance implies an exponential dichotomy. To do this it is convenient to use a non-euclidean norm on solutions. Let $\|x\| = \max_i |x_i|$ for the purposes of this section. We assume row dominance on A and that x is a non-trivial solution of $x' = A(t)x$.

Lemma 7.11. The function $\|x(t)\|$ has no relative maximum.

Proof: If $\|x(t)\|$ has a relative maximum at t_0, let $\|x(t_0)\| = |x_i(t_0)|$. Now $|x_i(t_0)|^2$ also has a relative maximum at t_0. Thus $x_i(t_0) \neq 0$ and

$$0 = \frac{1}{2}\left|\frac{d}{dt}|x_i(t_0)|^2\right| = \left|\mathrm{Re}(\overline{x_i(t_0)}\frac{d}{dt}x_i(t_0))\right|$$

$$= \left|\mathrm{Re}(\overline{x_i(t_0)}\{a_{ii}(t_0)x_i(t_0) + \sum_{j\neq i} a_{ij}(t_0)x_j(t_0)\})\right|$$

$$= \left|\mathrm{Re}(a_{ii}(t_0))|x_i(t_0)|^2 + \mathrm{Re}(\sum_{j\neq i} a_{ij}(t_0)x_j(t_0)\overline{x_i(t_0)})\right|$$

$$\geq |\mathrm{Re}\, a_{ii}(t_0)|\,|x_i(t_0)|^2 - \sum_{j\neq i}|a_{ij}(t_0)|\,|x_j(t_0)|\,|x_i(t_0)|$$

$$\geq \{|\mathrm{Re}\, a_{ii}(t_0)| - \sum_{j\neq i}|a_{ij}(t_0)\}\,|x_i(t_0)|^2 \geq \delta|x_i(t_0)|^2 > 0.$$

This contradiction proves the assertion.

Lemma 7.12. Suppose $\|x(t_0)\| = |x_i(t_0)|$ and $\mathrm{Re}(a_{ii}) > 0$ for all t. Then $\|x\|$ is strictly increasing on $[t_0,\infty)$.

Proof: First $x_i(t_0) \neq 0$ since $x \not\equiv 0$. By the computation in the preceding lemma, $\frac{1}{2}\frac{d}{dt}|x_i(t_0)|^2 \geq \delta|x_i(t_0)|^2 > 0$. For ρ small and positive $|x_i(t_0)| \leq |x_i(t_0+\rho)|$ and thus $\|x(t_0)\| = |x_i(t_0)| < |x_i(t_0+\rho)| \leq \|x(t_0+\rho)\|$. Consequently, $\|x\|$ is increasing at t_0. If there is a $t_1 \in [t_0+\rho,\infty)$ such that $\|x(t_1)\| \leq \|x(t_0+\rho)\|$, then $\|x\|$ has a relative maximum somewhere. This is not possible by the preceding lemma. If for $t_0+\rho < t_1 < t_2$

$\|x(t_2)\| \leq \|x(t_1)\|$, then on $(t_0+\rho, t_2)$ there is a relative maximum of $\|x\|$, again contradicting Lemma (7.11). Thus $\|x\|$ is strictly increasing on $(t_0+\rho, \infty)$ for all small ρ. Thus the lemma holds.

Lemma 7.13. If $\|x(t)\|$ is strictly increasing on R, then for every $t_0 \in R$ there is an i such that $\|x(t_0)\| = |x_i(t_0)|$ and $\operatorname{Re} a_{ii}(t) > 0$ for all t.

Proof: If not, let t_0 be such that $\|x(t_0)\| = |x_i(t_0)|$ implies $\operatorname{Re} a_{ii}(t_0) < 0$. For this t_0, divide the integers $1,\ldots,n$ into two classes so that $j \in I$ if and only if $\|x(t_0)\| = |x_j(t_0)|$ and $j \in J$ otherwise. There is an $\epsilon > 0$ so that $|x_j(t)| < \|x(t_0)\|$ if $j \in J$ and $t \in (t_0, t_0+\epsilon)$. Then for $i \in I$

$$\frac{1}{2}\frac{d}{dt}|x_i(t_0)|^2 = \operatorname{Re}(a_{ii}(t_0))\,|x_i(t_0)|^2$$

$$+ \operatorname{Re} \sum_{j \neq i} a_{ij}\, x_j(t_0)\overline{x_i(t_0)} \leq$$

$$\{\operatorname{Re} a_{ii}(t_0) + \sum_{j \neq i} a_{ij}\}\|x(t_0)\|^2 \leq -\delta\|x_0(t_0)\|^2 < 0.$$

There is an $\epsilon_2 > 0$ for which $|x_i(t)| < |x_i(t_0)|$ on $(t_0, t_0+\epsilon_2)$ and $i \in I$. On $(t_0, t_0+\min(\epsilon,\epsilon_2))$, $\|x(t)\| \leq \max_i |x_i(t)| < \|x(t_0)\|$ which is a contradiction.

Lemma 7.14. Suppose x is a solution such that $\|x\|$ is strictly increasing and δ is the constant in the definition of row dominance. Then for all $t_1 \leq t_2$, $\|x(t_1)\|e^{\delta(t_2-t_1)} \leq \|x(t_2)\|$.

Proof: If s is a given real number, then we can find an i such that $\|x(s)\| = |x_i(s)|$ and $\operatorname{Re} a_{ii}(t) > 0$ for all t. As before $\frac{d}{dt}|x_i(s)|^2 \geq 2\delta|x_i(s)|^2$ so, for $0 < \alpha < \delta$ there is an number $\rho(s,\alpha) > 0$ such that $x_i(t) \neq 0$ on $[s,s+\rho(s,\alpha)]$ and $\frac{d}{dt}|x_i(t)|^2 \geq 2\alpha|x_i(t)|^2$ there. It follows that $|x_i(t)|^2 \geq |x_i(s)|^2 e^{2\alpha(t-s)}$ if $t \in [s,s+\rho(s,\alpha)]$. This implies that $\|x(t)\|^2 \geq \|x(s)\|^2 e^{2\alpha(t-s)}$ on the same interval. So locally the Lemma is true. If t_1 is now an arbitrary real number, define

$$A \equiv \{\tau \mid \tau \geq t_1 \text{ and } \|x(t_1)\|^2 e^{2\alpha(t-t_1)} \leq \|x(t)\|^2 \text{ for } t_1 \leq t \leq \tau\}.$$

$A \neq \phi$ by the above. Suppose A is bounded above and let c be the least upper bound. Then by continuity $\|x(t_1)\|^2 e^{2\alpha(t-t_1)} \leq \|x(t)\|^2$ on $[t_1,c]$. Now for $t \in [c,c+\rho(c,\alpha)]$ we have

$$\|x(t)\|^2 \geq e^{2\alpha(t-c)}\|x(c)\|^2 \geq e^{2\alpha(t-c)} e^{2\alpha(c-t_1)} \|x(t_1)\|^2$$
$$= e^{2\alpha(t-t_1)} \|x(t_1)\|^2$$

so that $c+\rho(c,\alpha) \in A$. Thus $A = [t_1,\infty)$. Thus $t_1 \leq t_2$ implies that $\|x(t_1)\|^2 e^{2\alpha(t_2-t_1)} \leq \|x(t_2)\|^2$. Now $\alpha \in (0,\delta)$ is arbitrary so this holds for $\alpha = \delta$.

We now assume that the matrix A has $\operatorname{Re} a_{ii}(t) > 0$ for all t for exactly k subscripts. We still are assuming row dominance for all rows.

Lemma 7.15. For every positive integer m, there is a k-dimensional vector space $V_m \subset E^n$ such that if x is a nontrivial solution of $x' = Ax$ with $x(0) \in V_m$, then $\|x\|$ is strictly increasing on $[-m,\infty)$.

Proof: Let $S \equiv \{y \mid y_i = 0$ if $\operatorname{Re} a_{ii}(t) < 0$ for all $t\}$. Then S is k-dimensional. If $x(-m) \in S$ for a non-trivial solution, then by Lemma 7.12, $\|x(t)\|$ is strictly increasing on $[-m,\infty)$. If $X(t,\tau)$ is the fundamental solution of $x' = A(t)x$ with $X(\tau,\tau) = I$, then $V_m = \{c \mid c = X(0,-m)a,\ a \in S\}$ is a k-dimensional subspace since $X(0,-m)$ is non-singular.

We are now in a position to give the exponential dichotomy theorem.

Theorem 7.16. Suppose $A(t)$ satisfies the row dominance condition with parameter δ and that there are k subscripts i such that $\operatorname{Re} a_{ii}(t) > 0$ for all t. Then

a) there are k independent solutions $x_1,\ldots,x_k$ such that if $x \in \operatorname{span}(x_1,\ldots,x_k)$ then $\|x\|$ is strictly increasing and $\|x(t_1)\| e^{\delta(t_2-t_1)} \le \|x(t_2)\|$ if $t_1 \le t_2$.

b) there are $n-k$ independent solutions $y_1,\ldots,y_{n-k}$ such that if $x \in \operatorname{span}(y_1,\ldots,y_{n-k})$ then $\|x\|$ is strictly decreasing and $\|x(t_2)\| \le e^{-\delta(t_2-t_1)} \|x(t_1)\|$ if $t_1 \le t_2$.

c) The solutions $x_1,\ldots,x_k,y_1,\ldots,y_{n-k}$ are a fundamental set.

Proof: Define V_m as in the previous lemma. Let c_{mj} $j = 1,\ldots,k$ be an orthonormal basis in V_m. By the compactness of the unit sphere, we can extract a sequence of integers m_q so that $c_{m_q j} \to c_{oj}$ as $q \to \infty$ for $j = 1,\ldots,k$. Define the solutions x_j by $x_j(0) = c_{oj}$ $j = 1,\ldots,k$. Since c_{oj} are orthonormal, the x_j are independent. If $x \in \text{span}(x_1,\ldots,x_k)$ and $t_1 < t_2$, let the solution x_q satisfy the condition $x_q(0) = \sum_{i=1}^{k} \alpha_i c_{m_i}$ where $x = \sum_{i=1}^{k} \alpha_i x_i$. Then $\|x_q\|$ is strictly increasing on $[-m_q,\infty]$ so that $\|x_q(t_1)\| < \|x_q(t_2)\|$ if q is sufficiently large. On the compact interval $[t_1,t_2]$, $x_q \to x$ since the initial conditions of x_q tend toward the initial conditions of x. Thus $\|x(t_1)\| \leq \|x(t_2)\|$. Now t_1 and t_2 are arbitrary so that $\|x\|$ is nondecreasing. If it is not strictly increasing, then there is an interval on which it is constant. This cannot happen by Lemma 7.11. The exponential estimate follows from Lemma 7.14. This completes the proof of (a). The proof of part (b) is either done in a similar manner or one can make the change of equation, letting $B(t) = -A(-t)$ in which case one gets the hypotheses of this theorem with $n-k$ replacing k. Applying part (a) to this and reversing the time axis proves (b). The proof of part (c) is easy. For if $0 \not\equiv x \in \text{span}(x_1,\ldots,x_k) \cap \text{span}(y_1,\ldots,y_{n-k})$ then $\|x\| \to +\infty$ as $t \to \infty$ since $x \in \text{span}(x_1,\ldots,x_k)$. On the other hand $\|x\| \to 0$ as $t \to \infty$ since $x \in \text{span}(y_1,\ldots,y_{n-k})$. This is impossible so $\text{span}(x_1,\ldots,x_k) \cap \text{span}(y_1,\ldots,y_{n-k}) = \{0\}$ and $\text{span}(x_1,\ldots,x_k,y_1,\ldots,y_{n-k})$ is the vector space all solutions of $x' = A(t)x$ since its dimension is n.

Note that all vector norms are equivalent so that the exponential estimates of the theorem are true for every norm but there is a need to put a constant K in the formula, e.g. $\|x(t)\| \leq K\, e^{\delta(t-s)} \|x(s)\|$ if $s \leq t$ and $x \in \text{span}(y_1,\ldots,y_{n-k})$.

6. <u>Column dominant systems</u>. For this section we will use a different norm for vectors, namely the ℓ_1 norm, $\|x\| = \sum_{i=1}^{n} |x_i|$. We define a function $\sigma(f)(t)$ in the following way. The function f is supposed to be a real valued scalar function. Then

$$\sigma(f)(t) = \begin{cases} 1 & \text{if } f(t) > 0; \text{ or if } f(t) = 0 \text{ and } f'(t) > 0; \\ 0 & \text{if } f(t) = 0 \text{ and } f'(t) = 0; \\ -1 & \text{if } f(t) < 0; \text{ or if } f(t) = 0 \text{ and } f'(t) < 0. \end{cases} \tag{7.3}$$

Then it is easily verified that

$|f(t)| = f(t)\sigma(f)(t)$ and $D^+|f(t)| = f'(t)\sigma(f)(t)$ where D^+ denotes a right hand Dini derivate.

Now suppose that A is real and satisfies the column dominance condition with parameter δ, and $a_{ii}(t) < 0$ for all t. If x is a solution to $x' = A(t)x$, let $\sum$ be a diagonal matrix with $\sum_{ii}(t) = \sigma(x_i)(t)$. Furthermore, let 1 be the vector all of whose components are 1.

<u>Lemma 7.17</u>. With the above hypotheses $1^T \sum Ax \leq -\delta\|x\|$ and thus $D^+\|x\| \leq -\delta\|x\|$.

<u>Proof</u>: Note that $D^+\|x\| \leq \sum_{i=1}^{n} D^+|x_i| = \sum_{i=1}^{n} \sigma(x_i)(t)\ x_i'(t)$

$= 1^T \sum x' = 1^T \sum Ax$ so that the second estimate follows from the

first. Now $1^T \sum Ax = \sum_{i=1}^{n} \left(\sum_{j=1}^{n} a_{ij}(t)x_j(t) \right) \sigma(x_i)(t)$

$$= \sum_{j=1}^{n} [a_{jj} x_j(t)\sigma(x_j)(t) + \sum_{i\neq j} a_{ij}(t)x_j(t)\sigma(x_i)(t)]$$

$$= \sum_{j=1}^{n} [a_{jj}(t)|x_j(t)| + \sum_{i\neq j} a_{ij}(t)x_j(t)\sigma(x_i)(t)]$$

$$\leq \sum_{j=1}^{n} [a_{jj}(t)|x_j(t)| + \sum_{i\neq j} |a_{ij}(t)||x_j(t)|] \leq \sum_{j=1}^{n}(-\delta)|x_j(t)| = -\delta\|x\|.$$

Now it is well known that if f is a scalar function such that $D^+f(t) \leq \delta f(t)$, then $f(t) \leq y(t)$ for $t \geq t_0$ where y is the solution to $y' = -\delta y$ $y(t_0) = f(t_0)$. Applying this to $\|x\|$ we get as a corollary of Lemma 6.34 that $\|x(t)\| \leq e^{-\delta(t-s)}\|x(s)\|$ if $t \geq s$. Since all norms are equivalent, we get a similar estimate $|x(t)| \leq K e^{-\delta(t-s)}|x(s)|$ for any norm. Thus we have an exponential dichotomy with projection $P = I$, and $\sigma_1 = \sigma$. If one uses the matrix norm which is maximum of the ℓ_1 norms of the columns, then in fact $K = 1$.

In terms of stability this result is simply that $x = 0$ is an exponentially stable solution of $x' = A(t)x$.

7. Kinematic and Approximate Similarity. The problem of Kinematic Similarity has given rise to a great amount of research. In this section we review some of the work in this direction, citing some results but giving no proofs. In this connection we define the following classes of matrices, all $n \times n$.

$M_n \equiv$ {bounded continuous matrix functions on R with entries in C}

$AP_n = \{A \in M_n \mid A$ is almost periodic}

$AT_n = \{A \in AP_n \mid A$ is upper triangular$\}$

$K_n = \{$constant matrices in $M_n\}$

$T_n = \{$upper triangular matrices in $M_n\}$

$R_n = \{$real valued matrices in $M_n\}$

$AR_n = AP_n \cap R_n$.

The superscript $+$ means that the functions are only defined on $R^+ = [0,\infty)$. Also we write $A \overset{C}{\sim} B$ for A is kinematically similar to B with $P \in C$, where C is one of the above classes.

The historical development and definition of kinematic similarity may be found in Markus, L., Continuous matrices and the stability of differentiable systems, Math. Zeitsch. 62(1955), 310-319. He refers to the papers of Perron, Cameron, Lefschetz, and Diliberto.

Perron, O. Über eine Matrix-transformation. Math. Zeitzch. 32(1930), 465-473.

Cameron, R. H., Linear differential equations with almost periodic coefficients, Acta Math. Sturkh. 69(1938), 21-56.

Lefschetz, S. Lectures on differential equations, Princeton 1946.

Dilberto, S. On systems of ordinary differential equations, Contributions to the theory of nonlinear oscillations, Annal Studies 20, 1-39. Some of the results of Markus are the following.

<u>Theorem</u>. Let $A \in M_n^+$ and A commutes with $\int_0^t A$ for each t and $A = A_0 + A_p$ where A_0 is a constant. If A_0 commutes with A and $\int_0^t A$ with $\int_0^t A_p \in M_n^+$ then $A \overset{M_n^+}{\sim} A_0 = \lim_{t\to\infty} \frac{1}{t} \int_0^t A(s)ds$.

<u>Corollary</u>: If $A \in AP_n^+$, and A, $\int_0^t A$ and A_0 all commute on R^+ with $A_0 = M(A)$, and if $\int_0^t (A-A_0) \in M_n^+$, then $A \overset{AP_n}{\sim} A_0$.

<u>Corollary</u>: If $A \in AP_n^+$ and A commutes with $\int_0^t A$ and A has absolutely convergent fourier series with exponents bounded away from zero, then $A \overset{AP_n^+}{\sim} 0$.

This last result has the interpretation that the fundamental solution is almost periodic.

A great deal of the work on kinematic similarity has been done by Lillo. He introduces several notions related to the problem in his papers; Lillo, J. Linear differential equations with almost periodic coefficients, Amer. J. Math. 81(1959) 37-45; and Continuous Matrices and the Stability theory of Differential systems, Math. Zeitsch. 73(1960) 45-58.

<u>Definition</u>. A solution x of $x' = Ax$ has characteristic exponent λ if $\overline{\lim_{t\to\infty}} \frac{\log|x(t)|}{t} = \lambda$ and has the strong characteristic exponent λ if $\overline{\lim}$ is replaced by $\lim$. If there are e_i independent solutions with λ as characteristic exponent, then e_i is the multiplicity of λ.

Lillo remarks that the papers cited above by Markus and the paper of Markus suggests that problem of studying characteristic exponents of systems whose coefficient matrix in M_n^+ is equivalent to that for the coefficient matrix in T_n^+.

Theorem. If the diagonals entries $a_{ii}(t)$ have distinct mean values $\lambda_1,\dots,\lambda_n$, then the system $x' = Ax$ has $\lambda_1,\dots,\lambda_n$ as strong characteristic exponents.

Corollary. A system has distinct strong characteristic exponents $\lambda_1,\dots,\lambda_n$ if and only if it is kinematically similar to an upper triangular system whose diagonal entries have mean values $\lambda_1,\dots,\lambda_n$.

A further result may be found in Lillo's paper, Approximate similarity and almost periodic matrices, Proc. Amer. Math. Soc. 12(1961), 400-407.

Theorem. Suppose A has n distinct strong characteristic exponents and has n solutions x_i such that

$$M^{-1} e^{\lambda_j(t-t_0)} e^{-u|t-t_0|} \leq \frac{|x_j(t)|}{|x_j(t_0)|} \leq M e^{\lambda_j(t-t_0)} e^{u|t-t_0|} \quad \text{where}$$

$u = \min\limits_{i \neq j} \frac{|\lambda_i - \lambda_j|}{4}$. If $A \in AR_n$ then $A \overset{AR_n}{\sim} B \in AR_n \cap T_n$ and the diagonal elements of B have distinct mean values.

Lillo also introduces the notion of approximate kinematic similarity.

Definition. The matrix $A \in M_n^+$ is approximately kinematically similar to $B \in M_n^+$ if for every $\epsilon > 0$, there is a matrix $P(t,\epsilon)$ such that $P(t,\epsilon)$, $P^{-1}(t,\epsilon)$ and $P'(t,\epsilon)$ all are in M_n^+ and $|P^{-1}(-P' + AP) - B| < \epsilon$ on $[0,\infty)$. Equivalently A is approximately kinematically similar to B if there is a sequence $B_n \to B$ such that $A \sim B_n$. The idea behind this notion is the following. If one is interested in kinematic similarity from the viewpoint of stability theory, then if B is such that $x' = B(t)x$ has some stability property, then all near by equations should have the same property, and hence this can be transferred back to A.

Let $G_n \equiv \{A \in AR_n$: there is a matrix $B \in K_n$ with real distinct eigenvalues such that A is approximately kinematically similar to B by a member of $AR_n\}$. Lillo is able to show that if a system $x' = Ax$ has solutions x_i such that

$$\frac{1}{M} e^{\lambda_j(t-t_0)} e^{-u|t-t_0|} \leq \frac{|x_j(t)|}{|x_j(t_0)|} \leq M e^{\lambda_j(t-t_0)} e^{u|t-t_0|}$$

where M and λ_j are real numbers, λ_j distinct and $u = \min\limits_{i \neq j} \frac{|\lambda_i - \lambda_j|}{4}$, then $A \in G_n$. Furthermore, the set G_n is open in AR_n.

If one defines L_n in the same way as G_n except that B is not required to have real distinct eigenvalues, then it is not known if $AR_n \cap L_n = AR_n$, although this set is open in AR_n. He gives necessary and sufficient conditions for $A \in L_n$. These involve estimates on the solutions of the form cited immediately after the definition of G_n. It is of interest to note that if $A \in AR_n$ is approximately kinematically similar to B then so is every matrix in the hull of A.

Further results on kinematic similarity may be found in,

Langenhop, C., On bounded matrices and kinematic similarity, Trans.

Amer. Math. Soc. 97(1960), 317-326. He gives a necessary and sufficient condition for an element of M_n to be kinematically similar to an element of K_n by an element of M_n. It involves a condition on a generalized notion of eigenvalue which is related to the characteristic exponents.

8. Notes. The idea of exponential dichotomy and its consequences can be found in the book of Massera and Schaffer, Linear Differential Equations and Function Spaces, Academic Press, 1966. The book of Coppel [5] discusses the relationships to stability notions. Further, Coppel discusses these notions, some included here in Theorems 7.3-7.6, in his papers, Dichotomies and reducibility, J. Differential Equations, 4(1967), 500-521 and 4(1968), 386-398.

The notion of kinematic similarity is due to Markus and Example 7.8 is due to Markus and Yamake, Global stability criteria for differential systems, Osaka Math. J. 12(1960), 305-317.

Geschorgin's Theorem can be found in any advanced book on matrix theory. The notion of row dominance and the properties described in this chapter are in Lazer, J. Math. Anal. Appl. 35(1971), 215-229. Column dominance was exploited in Sandberg, Some theorems on the Dynamic response of nonlinear transistor networks, B.S.T.J. 48(1969), 35-54.

We have not proved the results in the section on kinematic and approximate similarity. The results are incomplete, in that they do not quite solve the intended problem. Nevertheless they are of great interest and a thorough exposition would require a great deal of space. The questions raised in this direction are very interesting and deserve more study.

Chapter 8

Fixed Point Methods

1. Introduction. A standard way to try to solve equations is through the use of fixed point methods. For differential and integral equations there are basically two methods, those that use some sort of contraction mapping and those that use the Schauder-Tychonov fixed point theorems. We will concern ourselves with elementary applications in this chapter. In a way we have a negative result, which says that the Schauder theorem is of little use in the context of almost periodic equations.

2. Applications of Contraction Mappings. We will be here considering systems that have an almost periodic linear part. The basic method is to look at an equation of the sort

$$x' = A(t)x + f(x,t) \tag{8.1}$$

with an almost periodic matrix A and where f is almost periodic in t uniformly for x in compact subsets of E^n. We select an almost periodic function φ and solve the equation

$$x' = A(t)x + f(\varphi(t),t) \tag{8.2}$$

in $AP(E^n)$, if possible. If this solution is $x = T\varphi$, then we are looking for a fixed point of the mapping T. To get the existence of $T\varphi$ we will need some sort of exponential dichotomy for the linear system

$$x' = A(t)x \quad . \tag{8.3}$$

Theorem 8.1. Suppose the system (8.3) satisfies an exponential dichotomy, where A is almost periodic. Further, let $f(x,t)$ be almost periodic in t uniformly for x in E^n and suppose it satisfies a Lipschitz condition, on E^n, of the form $|f(x,t) - f(y,t)| \le M|x-y|$ uniformly in t. There is a number L so that if $M < L$, then (8.1) has an almost periodic solution φ. Furthermore, $\mathrm{mod}(\varphi) \subset \mathrm{mod}(A,f)$.

Proof: We consider the complete metric space $B \equiv \{g \mid g \in AP(E^n)$ and $\mathrm{mod}(g) \subset \mathrm{mod}(f,A)\}$, using the uniform norm. Then if $\varphi \in B$, we have $f(\varphi(t),t) \in B$ by Theorem 4.6. We then use Theorem 7.7 to solve (8.2). Call this solution $T\varphi$. By that theorem, $\mathrm{mod}(T\varphi) \subset \mathrm{mod}(f(\varphi,t),A)$ so that $T\varphi \in B$. The mapping T satisfies by Theorem 7.7 (see representation in that proof)
$\|T\varphi - T\psi\| \le \left(\frac{K_1}{\sigma_1} + \frac{K_2}{\sigma_2}\right)\|f(\varphi(t),t) - f(\psi(t),t)\|$ and thus

$\|T\varphi - T\psi\| \le \left(\frac{K_1}{\sigma_1} + \frac{K_2}{\sigma_2}\right)M\|\varphi - \psi\|$. Let $L = \left(\frac{K_1}{\sigma_1} + \frac{K_2}{\sigma_2}\right)^{-1}$. If $M < L$ then T is a contraction mapping so that T has a fixed point.

The condition that f be almost periodic uniformly on E^n is somewhat severe. One can relax this slightly if some growth condition is placed on f. For example, by the estimate of Theorem 7.7 used above, T maps the ball $\|x\| \leq r$ into itself if $\left(\frac{K_1}{\sigma_1} + \frac{K_2}{\sigma_2}\right) g(r) \leq r$ where $g(r) = \sup_{\substack{t \in R \\ |x| \leq r}} |f(x,t)|$. Thus if $g(r) = o(r)$ as $r \to o$ or as $r \to \infty$, then this is satisfied. Then f need only be almost periodic uniformly on compact sets.

We also remark, that for A a constant matrix, the σ_i in the estimate for L are bounds for the eigenvalues, and in the diagonal dominant case are the σ of that definition. So if M is fixed, then $M < L$ if σ or σ_i are large enough. Note that the version of Theorem 5.11 where some of the eigenvalues of A are on the imaginary axis never applies here if f is non-periodic. This is because mod(f,A) will be dense in R if f or A is non-periodic.

3. <u>Applications of the Schauder Theorem</u>. We will try to formulate an approach to using the Schauder Theorem and see what we will need. If we look at the proof of Theorem 8.1, we see that we need the existence of $T\varphi$, so we will need an exponential dichotomy. To apply the Schauder Theorem we need a self map of some compact set. Now we will need a bounded set. But if $\|\varphi\| \leq r$ on which we define T, then $\|\varphi'\|$ is bounded by $\left(\frac{K_1}{\sigma_1} + \frac{K_2}{\sigma_2}\right) r$ and so the range of φ will consist of functions which satisfy a uniform Lipschitz condition. So the range of T restricted to a ball is a set of functions which satisfy the hypothesis of Corollary 4.10 except for

the exponent condition. Let us formulate this more precisely. Let $C(M,K,\Lambda) \equiv \{\varphi \mid \varphi \in AP(E^n),\ \|\varphi\| \leq M,\ |\varphi(t) - \varphi(s)| \leq K|t-s|,$ $\exp(\varphi) \subset \Lambda\}$, where M and K are given positive real numbers, and Λ is a given countable set of real numbers. According to Corollary 4.10, $C(M,K,\Lambda)$ is compact if and only if Λ has no finite limit point. Since we have $T\varphi$ defined on a space where $\exp f(\varphi)$ are contained in a subgroup Λ of the reals, then Λ has no finite limit points if and only if f is periodic. Now the periodic case does not require our heavy machinery because the functions could be restricted to a period and one then uses the Arzela-Ascoli Theorem to get compactness. So we look at the non-periodic case and conclude that it does not work.

This is not quite the case. It turns out that in certain circumstances, Theorem 4.6 can be improved to the semi-module.

<u>Theorem 8.2</u>. Let $f(z,t)$ be almost periodic in t uniformly for compact subsets of C^n. If f is an analytic function of z in a set K and φ is in $AP(C^n)$ with $\exp(\varphi) \subset$ semi-mod(f), then $\exp(f(\varphi(t),t)) \subset$ semi-mod(f).

<u>Proof</u>: We will show that $f(\varphi(t),t)$, which we know to be almost periodic, to be uniformly approximated by trigonometric polynomials with exponents in semi-mod(f).

The argument goes in the following way. We can approximate $f(z,t)$ uniformly by $\sum_{n=1}^{N} a(z,\lambda_n)e^{i\lambda_n t}$ on $K \times R$ with $\lambda_n \in \exp(f)$

and $a(z,\lambda_n)$ are fourier coefficients of f modified by constants independent of z. In fact, use the Bochner-Fejer sums. These constants, we noted, depended only on $\exp(f)$ and not on f or $a(f,x)$! It follows from Corollary 3.20 that $a(z,\lambda_n)$ are analytic in z on K also.

The second step in the argument now is that the sum $\sum_{n=1}^{N} a(z,\lambda_n)e^{i\lambda_n t}$ can be uniformly approximated on $K \times R$ by a sum of the form $\sum_{n=1}^{N} p_n(z)e^{i\lambda_n t}$ where $p_n(z)$ are polynomials. This is the use of the analyticity.

Now there is a $\delta > 0$ so that $|p_n(z) - p_n(y)| < \epsilon/N$ if $|z-y| < \delta$, $z,y \in K$ and $n = 1,\ldots,N$. We can uniformly approximate φ by $\sum_{n=1}^{N} b_n e^{i\mu_n t}$ with $\mu_n \in$ semi-mod(f) so that

$|\varphi(t) - \sum_{n=1}^{M} b_n e^{i\mu_n t}| < \delta$. It follows that $|p_n(\varphi(t)) - p_n(\sum_{n=1}^{M} b_n e^{i\mu_n t})| < \delta$

and so $\|\sum_{n=1}^{N} p_n(\varphi(t))e^{i\lambda_n t} - \sum_{n=1}^{N} p_n(\sum_{n=1}^{M} b_n e^{i\mu_n t})e^{i\lambda_n t}\| < c$. But $\sum_{n=1}^{N} p_n(\varphi(t))e^{i\lambda_n t}$ uniformly approximates $f(\varphi(t),t)$ and so $\sigma = \sum_{n=1}^{N} p_n(\sum_{n=1}^{M} b_n e^{i\mu_n t})e^{i\lambda_n t}$ does also. Since p_n is a polynomial, $p_n(\sum_{n=1}^{M} b_n e^{i\mu_n t})$ is of the form $\sum_{n=1}^{\overline{M}} c_n e^{i\nu_n t}$ where ν_n is a finite sum of the μ_n's with positive integral coefficients. Thus

$\nu_n \in$ semi-mod(f) and so $\exp(\sigma) \subset$ semi-mod(f).

To return to our discussion of the use of Schauder's Theorem, we see now that $C(M,K,\Lambda)$ can be compact if Λ = semi-mod(f) has no finite limit point. This requires that exp(f) have no finite limit point and all be of one sign. Note that all exponents being of one sign requires f to be complex valued. Finally note that the general exponential dichotomy does not preserve exponents but only modules. At least our proof only shows this. But to complete the above arguments we need to maintain that $\exp(T\varphi) \subset$ semi-mod$(f(\varphi,t),A)$. Thus we will require that A be constant. It seems that to apply the Schauder theorem is very hard. Note that adding smoothness conditions to $C(M,K,\Lambda)$ is of no help since the proof of Corollary 4.10 shows that there are too many entire functions in $C(M,K,\Lambda)$ if Λ has a finite limit point.

Theorem 8.3. Let $f(z,t)$ be almost periodic in t uniformly for z in compact subsets of C^n and suppose f is analytic in z in a region of C^n containing the origin so that $g(r) = \sup_{\substack{|z|\leq r \\ t\in R}} |f(z,t)|$ is $o(r)$ as $r \to o$. Let exp(f) all be one sign with no finite limit point and suppose the eigenvalues of A are disjoint from i semi-mod(f). Then there is an almost periodic solution to (8.1) with exponents in semi-mod(f).

Proof: Since the eigenvalues of A are not in i semi-mod(f) and this set has no finite limit point, the eigenvalues are positive distance from i semi-mod(f). According to Theorem 5.11 we can solve uniquely, the equation $x' = Ax + f(\varphi(t),t)$ if $\exp(\varphi) \subset$ semi-mod(f) and $x = T\varphi$ has $\exp(x) = \exp f(\varphi(t),t)$. According to that same theorem, the mapping T satisfies $\|T\varphi\| \leq M \|f(\varphi(t),t)\|$ for some M. We let $S \equiv \{\varphi \mid \varphi \in AP(C^n)\ \exp(\varphi) \subset \text{semi-mod}(f),\ \|\varphi\| \leq M_1,\ |\varphi(t) - \varphi(s)| \leq K|t-s|\}$ where we select M_1 in such a way that $Mg(M_1) \leq M_1$. Then $\|\varphi\| \leq M_1$ implies that $\|T\varphi\| \leq M_1$. Fixing M_1, f satisfies a Lipschitz condition on $|z| \leq r$ so that T is certainly continuous. Furthermore, $\|\varphi'\| \leq \|A\| M_1 + g(M_1)$ if $\|\varphi\| \leq M_1$. If we define $K = \|A\| M_1 + g(M_1)$ then T maps S into itself. Now S is a compact convex subset of $AP(C^n)$ so by the Schauder fixed point theorem, T has the required fixed point.

Note that in the situation given in the theorem, we do not get a result much different from Theorem 8.1, since under the hypothesis, if M_1 is small enough, the Lipschitz constant for f is small so that the map T turns out to be a contraction on a possibly smaller set than S. The results of this section then give evidence that the Schauder Theorem is inappropriate for our purposes.

4. A perturbation result. The results of the previous section can probably best be presented in the form of perturbation theorems. In this form we are going to look at the equation

$$x' = A(t)x + \epsilon\, g(x,t,\epsilon) \tag{8.4}$$

with the hypothesis that g is almost periodic in t uniformly for x in compact subsets of E^n and for each fixed small real ϵ. We de not require that the almost periodicity be uniform in ϵ, only that $g(x,t,\epsilon)$ be uniformly bounded on sets of the form $K \times R \times [0,\epsilon_0]$ where K is compact. There are two versions of the perturbation result.

<u>Theorem 8.4</u>. Let g satisfy the above hypothesis and satisfy a Lipschitz condition of the form $|g(x,t,\epsilon) - g(y,t,\epsilon)| \leq M(\epsilon_0)|x-y|$ where (x,t,ϵ) and (y,t,ϵ) are in sets of the form $K \times R \times [0,\epsilon_0]$. Suppose further that (8.3) satisfies an exponential dichotomy, A being almost periodic. Let $r > 0$. Then there is an $\epsilon_0 > 0$ so that if $0 \leq \epsilon \leq \epsilon_0$, then (8.4) has a unique almost periodic solution $x(t,\epsilon)$ with $\|x(t,\epsilon)\| \leq r$ and $\exp x(t,\epsilon) \subset \mathrm{mod}(A,g)$. Furthermore, $x(t,\epsilon)$ is continuous in ϵ with $x(t,0) \equiv 0$ if $g(x,t,\epsilon)$ is uniformly continuous in (x,t,ϵ) on $K \times R \times [0,\epsilon_0]$.

<u>Proof</u>: We find the norm M_1 of the mapping $\varphi \to T_1\varphi$ given by: $T_1\varphi$ is the almost periodic solution of $x' = A(t)x + \varphi$. Let $K = K(r) = \{x|\ |x| \leq r\}$. Pick ϵ so that $M_1\epsilon_1 M(\epsilon_1) < 1$. Note that $M(\epsilon)$ may be chosen to be a non-decreasing function of ϵ if the set K is fixed so that $M_1\epsilon M(\epsilon) < 1$ if $0 \leq \epsilon \leq \epsilon_1$. Let $|g(x,t,\epsilon)| \leq M_2$ for $|x| \leq r$, $t \in R$, and $0 \leq \epsilon \leq \epsilon_1$. Finally select ϵ_0 small so that $\epsilon_0 \leq \epsilon_1$ and such that $M_1\epsilon_1 M_2 \leq r$. With these definitions let $0 \leq \epsilon \leq \epsilon_0$,

$B \equiv \{\varphi \mid \varphi \in AP(E^n),\ \exp(\varphi) \subset \mathrm{mod}(A,f),\ \|\varphi\| \le r\}$; and $T\varphi$ be the almost periodic solution to $x' = A(t)x + \epsilon\, g(\varphi(t),t,\epsilon)$. Then $T\varphi = T_1 \epsilon\, g(\varphi(t),t,\epsilon)$ so $\|T\varphi\| \le M_1 \epsilon \|g(\varphi(t),t,\epsilon)\| \le M_1 \epsilon M_2 \le r$ if $\|\varphi\| \le r$. Thus T maps B into itself. Furthermore, $\|T\varphi - T\psi\| \le M_1 \epsilon \|g(\varphi,t,\epsilon) - g(\psi,t,\epsilon)\| \le M_1 \epsilon M(\epsilon_1) \|\varphi-\psi\|$ so that T is a contraction on the complete metric space B. The unique fixed point gives $x(t,\epsilon)$. Clearly $x(t,0) \equiv 0$. For the continuity of $x(t,\epsilon)$ in ϵ merely observe that the family $\{x(t,\epsilon):\ 0 \le \epsilon \le \epsilon_0\}$ is u.a.p. Since for $0 \le \epsilon \le \epsilon_0$

$$|x(t+\tau,\epsilon) - x(t,\epsilon)| = |T_1 \epsilon\, g(x(t+\tau),t+\tau,\epsilon) - T_1 \epsilon\, G(x(t),t,\epsilon)|$$
$$\le M_1 \epsilon\, |g(x(t+\tau),t+\tau,\epsilon) - g(x(t),t,\epsilon)|$$
$$\le M_1 \epsilon\, \{|g(x(t+\tau),t+\tau,\epsilon) - g(x(t),t+\tau,\epsilon)| + |g(x(t),t+\tau,\epsilon) - g(x(t),t,\epsilon)|$$
$$\le M_1 \epsilon_0 M(\epsilon_0) \|x(t+\tau) - x(t)\| + M_1 \epsilon_0 \sup_{|x| \le r} \|g(x_1 t+\tau,\epsilon) - g(x,t,\epsilon)\|$$

thus

$$\|x(t+\tau) - x(t)\| \le [1 - M_1 \epsilon_0 M(\epsilon_0)]^{-1} M_1 \epsilon_0 \sup_{\substack{|x| \le r \\ 0 \le \epsilon \le \epsilon_0}} \|g(x,t+\tau,\epsilon) - g(x,t,\epsilon)\|.$$

Now from the uniform continuity of g, its almost periodicity is uniform on $|x| \le r$ and $0 \le \epsilon \le \epsilon_0$. It follows that $v(\tau) = \sup_{\substack{|x| \le r \\ 0 \le \epsilon \le \epsilon_0}} \|g(x,t+\tau,\epsilon) - g(x,t,\epsilon)\|$ is almost periodic. Thus $\{x(t,\epsilon):\ 0 \le \epsilon \le \epsilon_0\}$ is u.a.p. If $\epsilon_n \to \tilde{\epsilon}$, then $x(t,\epsilon_n) \to \varphi(t)$ uniformly for some subsequence. Clearly φ is an almost periodic solution of $x' = A(t)x + \tilde{\epsilon}\, g(x,t,\tilde{\epsilon})$ and is in B so it must be $x(t,\tilde{\epsilon})$. This gives the continuity.

We also do the complex version of this theorem since it gives possible different estimates on ϵ_0.

Theorem 8.5. Let $g(z,t,\epsilon)$ be analytic in z near $z = 0$ and almost periodic in t for fixed ϵ uniformly on compact subsets of C^n. Let g be bounded on sets of the form $K \times R \times [0,\epsilon_0]$. Let $\exp(g)$ be all of one sign with no limit points and A have eigenvalues not in i semi-mod(g). Let $r > 0$. There exists ϵ_0 such that if $0 \leq \epsilon \leq \epsilon_0$, then there is an almost periodic solution φ to (8.4) with $\exp(\varphi) \subset$ semi-mod(f).

Proof: We set up the problem as in Theorem 8.2. Let $r > 0$ be given and let M_1 be the norm of the mapping $\varphi \to T_1\varphi$ where $T_1\varphi$ is the solution of $x' = Ax + \varphi$ with $\exp(T_1\varphi) \subset$ semi-mod$(\varphi) \subset$ semi-mod(g). Let ϵ_0 be such that

$$\epsilon_0 \sup_{\substack{|z|\leq r \\ 0\leq\epsilon\leq\epsilon_0}} |g(z,t,\epsilon)| M_1 \leq r$$

then if

$B \equiv \{\varphi \mid \varphi \in AP(C^n)$ $\exp(\varphi) \subset$ semi-mod(g), $\|\varphi\| \leq r$ and $|\varphi(t) - \varphi(s)| \leq K|t-s|\}$ we have $T\varphi = T_1 \epsilon g(\varphi,t,\epsilon)$ maps B into itself if $0 \leq \epsilon \leq \epsilon_0$ and $K = \|A\|r + \epsilon_0 \sup_{\substack{|z|\leq r \\ 0\leq\epsilon\leq\epsilon_0}} |g(z,t,\epsilon)|$. By the Schauder theorem there is a fixed point.

Since we do not have uniqueness, we cannot guarantee continuity of this solution in ϵ.

5. Applications to more general equations. Suppose we want to look at equations which don't have the ϵ parameter in them. Sometimes it is possible to introduce it. Let us take the situation

$$x' = A(t)x + f(x,t) + \epsilon^2 p(t) \tag{8.5}$$

where p is almost periodic and A is almost periodic. If $f(x,t) = O(|x|^2)$, as an example, and has sufficient continuity properties, then $f(\epsilon x,t) = \epsilon^2 g(t,x,\epsilon)$ for a function g with roughly the same continuity properties and the same almost periodicity properties as f. The change of variable $x = \epsilon y$, $\epsilon > 0$ reduces (8.5) to

$$y' = A(t)y + \epsilon\, g(x,t,\epsilon) + \epsilon\, p(t) \tag{8.6}$$

where now (8.6) is of the form (8.4). In particular, if $f(x,t)$ is almost periodic in t uniformly for x in compact sets and is analytic in x near 0, then $g(x,t,\epsilon)$ also has this property for ϵ small.

In particular, the scalar equation $x' = f(x) + \epsilon^2 p(t)$ with f analytic and p almost periodic and $f(x) = O(|x|^2)$ as $x \to 0$ is equivalent to the pair of real equations

$$\begin{aligned} \xi' &= u(\xi,\eta) + \epsilon^2 p_1(t) \\ \eta' &= v(\xi,\eta) + \epsilon^2 p_2(t) \end{aligned} \tag{8.7}$$

where u and v are harmonic conjugates and $p = p_1 + ip_2$. One specific example is $f(z) = z^2$ so that one has

$$\xi' = \xi^2 - \eta^2 + \epsilon^2 p_1(t)$$

$$\eta' = 2\xi\eta + \epsilon^2 p_2(t)$$

which now has almost periodic solutions for ϵ small.

6. Notes. The application of the contraction mapping is standard. A specific reference for Theorem 8.1 is either Bogdanowitz, On the existence of almost periodic solutions for systems of ordinary differential equations in Banach spaces, Arch. Rational Mech. Anal. 13(1963), 364-370 or Coppel, Almost periodic properties of ordinary differential equations, Annali di Math. pura ed appl. 76(1967) 27-50. This latter paper also has Theorem 8.2.

The results in Theorem 8.3 and following can be found in Fink, Compact families of almost periodic functions and an application of the Schauder fixed point theorem, SIAM J. Appl. Math. 17(1969), 1258-1262, and Fink and Seifert, Non-resonance conditions for the existence of almost periodic solutions of almost periodic systems, SIAM J. Appl. Math. 21(1971), 362-366.

Chapter 9

Asymptotic almost periodic functions and other weaker conditions

1. Introduction. Soon after the invention of almost periodic functions, or in fact, even earlier in the researches of Bohl, it was recognized that stability of solutions to differential equations was closely related to almost periodicity. In our exposition of these results, we will use a technique which first proves a condition that is weaker than almost periodicity. This condition, which we call asymptotic almost periodicity then implies the existence of an almost periodic solution.

2. Asymptotic almost periodicity and solutions.

Definition 9.1. Let φ be defined on R^+ to E^n. We suppose that φ is continuous. Then φ is asymptotically almost periodic if and only if there is an almost periodic function p and a continuous function q defined on R^+ with $\lim_{t\to\infty} q(t) = 0$ such that $\varphi = p + q$ on R^+. We abbreviate asymptotic almost periodicity by a.a.p. The function p is called the almost periodic part.

The main justification for the introduction of this concept is the following theorem.

Theorem 9.2. If $f(x,t)$ is almost periodic in t uniformly for x in compact subsets of E^n and φ is an a.a.p. solution of $x' = f(x,t)$ on R^+, then the almost periodic part is a solution on R.

Proof: Let $\varphi = p+q$ and let $\alpha'_n = n$. Since p is almost periodic we select a subsequence $\alpha \subset \alpha'$ so that $T_\alpha p = \psi$ and $T_\alpha f = g$ exists uniformly. Now $T_\alpha q = 0$ uniformly on compact sets. Thus $T_\alpha \varphi = \psi$. Note that as in the proof of Theorem 6.2, ψ is now a solution, on R, of the equation $x' = g(x,t)$ and is almost periodic. Now $T_{-\alpha} g = f$ and $T_{-\alpha}\psi = p$ uniformly on R so p is a solution of $x' = f(x,t)$ on R.

3. Alternate definitions. As in the almost periodic case, there are several versions of a.a.p. which are equivalent. Some of these are more useful than others in certain cases so we indicate proofs of most of them. We use the notations $\alpha > 0$ to mean $\alpha_n > 0$ for all n, and $T_\alpha(\mathrm{id}) = +\infty$ to mean $\lim_{n\to\infty} \alpha_n = +\infty$.

Theorem 9.3. With continuity as a standing hypothesis, any of the following conditions are equivalent to a.a.p. of f.

1) For every $\alpha' > 0$ with $T_{\alpha'}(\mathrm{id}) = +\infty$, there exist $\alpha \subset \alpha'$ so that for every $\beta \in R$, $T_\alpha f$ exists uniformly on $[\beta,\infty)$.

2) Condition (1) for $\beta = 0$ only.

3) For every $\epsilon > 0$, there is a $T(\epsilon)$ such that $\{\tau \mid \sup|f(t+\tau) - f(t)| < \epsilon$, sup over $t \geq T$, $t+\tau \geq T\}$ is relatively dense in R.

4) Condition (3) except the set need only be relatively dense in R^+.

5) Given $\alpha' > 0$ $T_{\alpha'}(\mathrm{id}) = +\infty$, there exists $\alpha \subset \alpha'$ and a number $d(\alpha) > 0$ so that $T_\alpha f$ exists pointwise on R^+

and if $\delta > 0$ $\beta \subset \alpha$, $r \subset \alpha$ are such that $T_{\delta+\beta} f = g$ and $T_{\delta+r} f = h$ exist pointwise on R^+ then either $g \equiv h$ or $|g(t) - h(t)| \geq 2d(\alpha)$ on R^+.

Proof: A.a.p. $\Rightarrow$ (1) It is sufficient to note that since $q(t) \to 0$ as $t \to \infty$ and $\alpha'_n \to \infty$, $T'_\alpha q = 0$ uniformly on compact sets. Thus invoke $\alpha \subset \alpha'$ so that $T_\alpha p$ exists uniformly.

(1) $\Rightarrow$ (4) If for some $\epsilon > 0$, no T exists we do the following. To start, take $T = 0$ and $\alpha_1 = 1$. There is an interval $(a_2, b_2) = I_2$ such that $b_2 - a_2 > 2$ and containing no τ such that $\sup_{\substack{t \geq 0 \\ t+\tau \geq 0}} |f(t+\tau) - f(t)| < \epsilon$. Let α_2 be the center of this interval.

Then $\alpha_2 - \alpha_1 \in I_2$ so is not a τ for R^+. There is an interval $I_3 = (a_3, b_3) \subset [\alpha_2, \infty)$ such that $b_3 - a_3 > 2(\alpha_1 + \alpha_2)$ and containing no τ for $[\alpha_1, \infty)$, i.e. no τ such that $\sup_{\substack{t \geq \alpha_2 \\ t+\tau \geq \alpha_2}} |f(t+\tau) - f(t)| < \epsilon$.

Note that $\alpha_3 - \alpha_2$ and $\alpha_2 - \alpha_1 \in I_3$. Repeat this; having selected intervals I_n such that length $I_n > 2(\sum_{i=1}^{n-1} \alpha_i)$ with α_i the center of I_i, take I_{n+1} with center α_{n+1} such that $I_{n+1} \subset [\alpha_n, \infty)$ and containing no τ for which $\sup_{\substack{t \geq \alpha_n \\ t+\tau \geq \alpha_n}} |f(t+\tau) - f(t)| < \epsilon$. Then $\alpha_{n+1} - \alpha_i \in I_{n+1}$ for $i = 1, \ldots, n$ so are not τ's on $[\alpha_n, \infty)$.

We consider the sequence α defined in this way. Clearly $\alpha > 0$ and $T_\alpha(\mathrm{id}) = +\infty$. Note that if $i < j$, then

$\sup_{x \geq 0} |f(x+\alpha_i) - f(x+\alpha_j)| = \sup_{y \geq \alpha_i} |f(u) - f(u+\alpha_j-\alpha_i)| \geq \epsilon$, since $\alpha_j - \alpha_i \subset I_j$ and $\alpha_i \leq \alpha_{j-1}$. This shows that for no subsequence $\alpha' \subset \alpha$ is it possible that $T_{\alpha'}f$ exists uniformly on R^+.

(4) $\Rightarrow$ (3) Let $\epsilon > 0$ be given and let $T(\epsilon)$ be given by (4) with ℓ the inclusion length on R^+. Let I be an interval of R of length at least ℓ. If $I \subset R^+$ it contains a τ by (4). If $0 \in I$, take $\tau = 0$, so we assume $I \subset (-\infty,0)$. Let $I^* = -I$. Let $\tau^* \in I^*$ which has the property that $|f(t+\tau^*) - f(t)| < \epsilon$ if $t \geq T$ and $t+\tau^* \geq T$. Let $\tau = -\tau^*$ and $\sigma = t+\tau^*$. Then $\tau \in I$. We have $|f(\sigma) - f(\sigma+\tau)| < \epsilon$ if $\sigma + \tau \geq T$ and $\sigma \geq T$ as was to be shown.

(3) $\Rightarrow$ a.a.p. Let $\alpha'_n = n$. We assert that f is bounded and uniformly continuous on R^+. The uniform continuity is an exact duplicate of Theorem 1.13 and the boundedness follows in the same way. Thus the family $\{f(t+\alpha'_n), n \geq 1\}$ is uniformly bounded and equi-uniformly continuous on R^+, so we let $\alpha \subset \alpha'$ so that $T_\alpha f = p$ exists uniformly on compact subsets of R. We first show that p is almost periodic. If $\epsilon > 0$ is given find T according to (3) and let ℓ be the inclusion length, i.e. every interval of length ℓ in R contains a τ such that $|f(t+\tau) - f(t)| < \epsilon$ if $t \geq T$ and $t+\tau \geq T$. Introducing α_n we have

$$|f(t+\alpha_n+\tau) - f(t+\alpha_n)| < \epsilon \quad \text{if} \quad t \geq T-\alpha_n \quad \text{and} \quad t+\tau \geq T-\alpha_n.$$

Fix t and τ, and take n large so the last two inequalities are correct. Then taking limits one has $|p(t+\tau) - p(t)| < \epsilon$. This holds for $t \in R$ and a τ in every interval of length ℓ. Thus p is in $AP(E^n)$. To complete the proof we will show that $f(t) - p(t) \to 0$ as $t \to \infty$. Let $\epsilon > 0$ be given and pick T as in (3) and $\tau_n \in [\alpha_n - \ell, \alpha_n]$. For large n, $\tau_n > 0$ so that $|f(t+\tau_n) - f(t)| < \epsilon$ if $t \geq T$. Here ℓ is the inclusion length. Define $\ell_n = \alpha_n - \tau_n$. Then $0 \leq \ell_n \leq \ell$ so that by taking subsequences, and not changing notation, we may assume $\ell_n \to \ell^*$. We have by the first part of the proof

$$|p(t) - f(t+\alpha_n)| < \epsilon \tag{9.1}$$

for n large and $t \geq T$. By uniform continuity,

$$|p(t-\ell_n) - p(t-\ell^*)| < \epsilon \tag{9.2}$$

for n large. Recall that $|f(t+\tau_n) - f(t)| < \epsilon$ if $t \geq T$. For large n, we may replace t by $t-\ell_n$ in (9.1) to get $|p(t-\ell_n) - f(t+\tau_n)| < \epsilon$. We combine this with (9.1) and (9.2) to get

$$|f(t) - p(t-\ell^*)| < |f(t) - f(t+\tau_n)| + |f(t+\tau_n) - p(t-\ell_n)|$$

$$+ |p(t-\ell_n) - p(t-\ell^*)| < 3\epsilon \qquad (9.3) \qquad \text{if } t \geq T.$$

This does not suffice to prove the theorem since ℓ^* depends on ϵ. To remedy this write $\ell_n^* = \ell^*(\frac{1}{n})$ when $\epsilon = 1/n$. There is a subsequence $k \subset \ell^*$ such that $T_{-k}p = p^*$ exists uniformly on R.

Claim that $\lim_{t\to\infty} [f(t) - p^*(t)] = 0$ so that $f(t) - p^*(t) = q(t)$ is the required decomposition. If η is given, pick $\epsilon = \eta/6$ in (9.3) to get $|f(t) - p(t-k_n)| < \eta/2$ if $n \geq N$, and $t \geq T(\eta/6)$. If n is large then $|p(t-k_n) - p^*(t)| < \eta/2$ so $|f(t) - p^*(t)| < \eta$ if $t \geq T(\eta/6)$, as required. Note that $p^* = p$ is necessarily true since the limit of $f(t)$ must be unique.

(a.a.p.) $\Rightarrow$ (5) Let α' be given with $T_{\alpha'}(id) = +\infty$. Let $\alpha \subset \alpha'$ so that $T_\alpha f = g$ uniformly on R^+. Take $d(\alpha) = 1$. Let $\delta > 0$ and $\beta \subset \alpha$ and $r \subset \alpha$ so that $T_{\delta+\beta} f = h_1$ and $T_{\delta+r} f = h_2$ exist pointwise. Since $T_{\alpha'}(id) = +\infty$ g is almost periodic and we may choose $\delta' \subset \delta$ so that $T_{\delta'} g$ exists uniformly on R. If β' and r' denote the subsequences of β and r that are common with δ' then for $f = p+q$, and $\lim_{t\to\infty} q(t) = 0$, we have $h_1 = T_{\delta'+\beta'} f$

$= T_{\delta'+\beta'} p + T_{\delta'+\beta'} q = T_{\delta'+\beta'} p = T_{\delta'} T_{\beta'} p = T_{\delta'} (T_\alpha f) = T_{\delta'} g.$

In a similar way, $h_2 = T_{\delta'} g$ so that $h_1 \equiv h_2$.

(5) $\Rightarrow$ (2) Let α' be such that $T_{\alpha'}(id) = +\infty$. Let $\alpha \subset \alpha'$ such that $T_\alpha f$ exists pointwise on R^+. If the convergence is not uniform, there are sequences $\delta' > 0$, $\beta' \subset \alpha$ $\gamma' \subset \alpha$ and $\epsilon > 0$ such that

$$|f(\beta_n'+\delta_n') - f(\gamma_n'+\delta_n')| \geq \epsilon$$

for all n. For this purpose we may take $\epsilon < d(\alpha)$. Since $T_\alpha f(0)$ exists we have $|f(\beta_n') - f(\gamma_n')| < d(\alpha)$ for large n. Thus the function $g(t) = f(t+\beta_n') - f(t+\gamma_n')$ satisfies $|g(0)| < d(\alpha)$ and

$|g(\delta_n')| \geq \epsilon$ for large n. There exists $\delta_n'' > 0$, for large n, so that $\epsilon \leq |g(\delta_n'')| \leq d(\alpha)$. Now define the sequences $\beta'' \subset \beta'$ and $\gamma'' \subset \gamma'$ common with $\delta'' \subset \delta'$. By first statement of (5), there are subsequences $\delta \subset \delta''$, $\beta \subset \beta''$, $\gamma \subset \gamma''$, all common; so that $T_{\beta+\delta}f = h_1$ and $T_{\gamma+\delta}f = h_2$ exist pointwise on R^+. Applying (5) we must have $h_1 \equiv h_2$ or $|h_1(t) - h_2(t)| \geq 2d(\alpha)$ on R^+. But $|h_1(0) - h_2(0)| = \lim_n |f(\delta_n+\beta_n) - f(\delta_n+\gamma_n)|$ is a number in $[\epsilon, d(\alpha)]$. This contradiction shows that $T_\alpha f$ exists uniformly on R^+.

(2) $\Rightarrow$ (1) Let α be given so that $T_\alpha(\text{id}) = +\infty$ and $T_\alpha f$ exists uniformly on R^+. We show this is uniform on $[\beta,\infty)$. In fact if β is given, $t+\alpha_n \geq \beta$ for n large. Let $\sigma = t-\beta$ then $\sigma \geq 0$ if and only if $t \geq \beta$, and $f(t+\alpha_n) = f(\sigma+\alpha_n+\beta)$. Now $\alpha_n+\beta \to +\infty$ so the right hand side exists uniformly on $\sigma \geq 0$ or $t \geq \beta$ as was to be shown.

<u>Corollary 9.4</u>. A function f is almost periodic if and only if from every α', one can extract $\alpha \subset \alpha'$ such that $T_\alpha f$ exists pointwise and there is a number $d(\alpha) > 0$ such that if $T_{\delta+\beta}f = h_1$ and $T_{\delta+\gamma}f = h_2$ exist pointwise for $\beta \subset \alpha$, $\gamma \subset \alpha$, then either $h_1 \equiv h_2$ or $|h_1(t) - h_2(t)| \geq 2d(\alpha)$.

<u>Proof</u>: Repeat the proof (5) $\Rightarrow$ (2) in above theorem without requiring that $\delta' > 0$, or that $T_{\alpha'}(\text{id}) = +\infty$, and one concludes the Bochner criterion. The reverse is simple since $h_1 = h_2$ by using the property $T_{\delta+\beta}f = T_\delta T_\beta f = T_\delta T_\alpha f = T_\delta T_\gamma f = T_{\delta+\gamma}f$.

4. <u>An application</u>. We will show how the concept of a.a.p. can be

used for differential equations without the need for almost periodicity in the equation.

Theorem 9.5. Let $x' = A(t)x$, A almost periodic, satisfy an exponential dichotomy with $\sigma_i = 0$, and suppose all bounded solutions of $x' = A(t)x + b(t)$ are almost periodic, here $b \in AP(E^n)$. Suppose $X(t)P \to 0$ as $t \to \infty$ where P is the projection and X the fundamental solution in the exponential dichotomy. Finally let $f(x,t)$ be continuous, $\int_0^\infty |f(0,t)|dt < \infty$ and $|f(x,t) - f(y,t)| \leq L(t)|x-y|$ with $\int_0^\infty L(t)dt < \infty$. Then there is a 1:1 correspondence between bounded solutions of $x' = A(t)x + b(t)$ and

$$x' = A(t)x + b(t) + f(x,t) . \tag{9.4}$$

Furthermore, the difference of the matched pairs goes to 0 as $t \to \infty$. That is, all solutions of (9.4) are a.a.p.

Proof: Pick t_0 large so that $\theta = \max(K_1,K_2)\int_0^\infty L(s)ds < 1$ where the K_i are the estimates from exponential dichotomy. For any $x \in C[t_0,\infty)$ define

$$Tx(t) = \int_{t_0}^t X(t)PX^{-1}(s)f(x(s),s)ds - \int_t^\infty X(t)(I-P)X^{-1}(s)f(x(s),s)ds.$$

Note that

$$|Tx(t)| \leq \int_{t_0}^t K_1\ f(x(s),s)ds + \int_t^\infty K_2\ f(x(s),s)ds$$

$$\le K\int_{t_0}^{\infty} [f(x(s),s) - f(0,s)] + [f(0,s)]ds$$

$$\le K\int_{t_0}^{\infty} L(s)x(s) + K\int_{t_0}^{\infty} |f(0,s)|ds < \infty, \; K = \max(K_1,K_2).$$

So T takes bounded functions into bounded functions. By a similar computation,

$$|Tx(t) - Ty(t)| \le K\int_{t_0}^{\infty} |f(x(s),s) - f(y(s),s)|ds$$

$$\le K\int_{t_0}^{\infty} L(s)ds \; \|x-y\| = \theta\|x-y\|$$

so that T is a contraction in the uniform norm. Thus if z is continuous and bounded $Sx \equiv z + Tx$ has a unique fixed point that is, $\bar{x} = z + T\bar{x}$. Now $\bar{x} - z = T\bar{x}$ is differentiable and $(\bar{x}-z)'(t) = f(\bar{x}(t),t) + A(\bar{x}-z)(t)$. If z is a bounded (hence almost periodic) solution of $x' = Ax + b$, then x is differentiable and x is a solution of (9.4). Conversely, if x is a solution of (9.4) and $z = x - Tx$, then z is differentiable and solves $x' = Ax + b$. This gives a 1:1 map $x \to z$. The continuity follows from $\|z_1 - z_2\| \le \|x_1 - x_2\| + \|Tx_1 - Tx_2\| \le (1+\theta)\|x_1 - x_2\|$ and $\|x_1 - x_2\| \le \|z_1 - z_2\| + \|Tx_1 - Tx_2\| \le \|z_1 - z_2\| + \theta\|x_1 - x_2\|$ so that $\|x_1 - x_2\| \le (1-\theta)^{-1}\|z_1 - z_2\|$. To get the above on $(0,\infty)$ instead of (t_0,∞) extend backwards by uniqueness of initial value problems. To get the asymptotic property, choose $t_1 > t_0$ so that

$K\int_{t_1}^{\infty}[L(s)(x(s))| + |f(0,s)|]ds < \epsilon$. Then for $t \geq t_1$, we have

$$|Tx(t)| \leq \int_{t_0}^{t} |X(t)PX^{-1}(s)|\,|f(x(s),s)|ds$$

$$+ K\int_{t}^{\infty} |f(x(s),s)|ds$$

$$\leq |X(t)P|\int_{t_0}^{t_1} |X^{-1}(s)f(x(s),s)|ds + \epsilon$$

Since $X(t)P \to 0$ as $t \to \infty$ we have $Tx(t) \to 0$ as $t \to \infty$ and x and z are asymptotic.

Note that the Lipschitz condition does not allow f to be in AP.

5. Extension of almost periodic sequences. As we have observed before, the notion of almost periodicity makes sense on any additive group. In particular along any arithmetic sequence in R. We take the integers Z. For the definition of an almost periodic sequence we may look at the Bohr definition and just restrict t to be in Z.

Let $\{a_n\}$ be an almost periodic sequence. We can define a function on R by $f(t) = s\,a_n+(1-s)a_{n+1}$ if $t = sn+(1-s)(n+1)$, $0 \leq s \leq 1$. Clearly $f(n) = a_n$ and f is continuous. It is easy to see that $f \in AP(E^n)$ if $a_m \in E^n$ for all m, since f is the linear extension of a_n.

Conversely, if f is in $AP(E^n)$, then $f(n)$ is an almost periodic sequence. This is not quite as easy to see. It is almost equivalent to the statement that for every $\epsilon > 0$, $Z \cap T(f,\epsilon)$ is relatively dense. In any case, the latter is sufficient and turns out to be true. To see this, consider the function g which is $(-1)^n$ at n and linear in between. It is easy to verify, that if ϵ is sufficiently small, then $\tau \in T(g,\epsilon)$ if and only if $|\tau-n| < \delta(\epsilon)$ for some integer n and a function $\delta(\epsilon)$ for which $\delta(\epsilon) \to 0$ as $\epsilon \to 0$. In fact δ is linear. Now if $\epsilon > 0$ is given, there is an η so that if $|s-\tau| < \eta$ and $\tau \in T(f,\epsilon/2)$, then $s \in T(f,\epsilon)$. In fact η comes from uniform continuity of f. Take ϵ^* so that $\delta(\epsilon^*) < \eta$, and $S = T(f,\epsilon/2) \cap T(g,\epsilon^*)$. If $\tau \in S$ then $|\tau-n| < \eta$ and $\tau \in T(f,\epsilon/2)$. Thus $n \in T(f,\epsilon)$. Now if ℓ is the inclusion length of S, then $(\ell+1)$ is an inclusion length for $T(f(n),\epsilon)$.

Thus we see that if f is an almost periodic solution to a differential equation, its restriction to an arithmetic sequence is an almost periodic sequence. Is the converse true? We have seen that filling in the gaps linearly gives an almost periodic function. Does the differential equation "fill it in" almost periodically? The answer is yes. We need a lemma.

Lemma 9.6. Let $f(x,t)$ be almost periodic in t uniformly on compact subsets of E^n. Suppose that every equation in the hull has unique solutions to initial value problems. For every $\epsilon > 0$, there is a $\delta > 0$ so that for $g_1(x,t)$ and $g_2(x,t)$ in $H(f)$ with $\sup_{x\in K}\|g_1(x,t) - g_2(x,t)\| < \delta$ and φ_i a solution of $x' = g_i(x,t)$ in K with $|\varphi_1(0) - \varphi_2(0)| < \delta$, $(\varphi_i(0) \in K)$, we also have $|\varphi_1(t) - \varphi_2(t)| < \epsilon$ for $0 \leq t \leq 1$.

Proof: If the lemma is not true, there is an $\epsilon > 0$ and $g_n^{(i)}(x,t) \in H(f)$ such that $\sup_{x\in K}\|g_n^{(1)}(x,t) - g_n^{(2)}(x,t)\| < 1/n$, and solutions $\varphi_n^{(i)}$ of $x' = g_n^{(i)}(x,t)$ with $|\varphi_n^{(1)}(0) - \varphi_n^{(2)}(0)| < \frac{1}{n}$ but $\sup_{0\le t\le 1} |\varphi_n^{(1)}(t) - \varphi_n^{(2)}(t)| \ge \epsilon$. Since $\varphi_n^i(0) \in K$ we may assume that $\varphi_n^{(i)}(0) \to x_0$, $i = 1,2$. Similarly we may assume that $g_n^i(x,t) \to g(x,t) \in H(f)$. Again by taking subsequence we may assume $\varphi_n^i(t) \to \psi^i(t)$ uniformly on compact subsets of R. By Kamke's theorem and unique solutions to initial value problems, ψ^i are solutions to $x' = g(x,t)$, $x(0) = x_0$. Thus $\psi^1 \equiv \psi^2$ contradicting $\sup_{0\le t\le 1} |\varphi_n^1(t) - \varphi_n^2(t)| \ge \epsilon$.

Theorem 9.7. Let $f(x,t)$ be almost periodic in t uniformly for x in compact subsets of E^n. Suppose further, that all equations $x' = g(x,t)$ with $g \in H(f)$ have unique solutions to initial value problems. If φ is a bounded solution, then φ is almost periodic if and only if its restriction to some discrete subgroup of the reals is an almost periodic sequence.

Proof: The restriction of an almost periodic function to a discrete subgroup is an almost periodic sequence as we observed above. It is the converse that is of interest here. By a change of scale we may assume that Z is the discrete subgroup in question. Let $\epsilon > 0$ be given and choose $\delta(\epsilon)$ as in Lemma 9.6. Let $T_1 = T(\varphi(n),\delta)$ and $T_2 = T(f,\delta)$. Since the linear extension of φ

to R is almost periodic, call it $\overline{\varphi}$, we have $T(\overline{\varphi},\delta) \cap T_2$ has a relatively dense set of integers which are now in $T_1 \cap T_2$. If $m \in T_1 \cap T_2$, then $|f(x,t+m) - f(x,t)| < \delta$, $x \in \text{Range}(\varphi)$, all t, and $|\varphi(n+m) - \varphi(n)| < \delta$ for all n. Let $n \leq t \leq n+1$ and consider $\psi_1(t) = \varphi(t+n+m)$ and $\psi_2(t) = \varphi(t+n)$. The function ψ_1 is a solution of $x' = f(x,t+n+m)$, $\psi_1(0) = \varphi(n+m)$ and ψ_2 is a solution to $x' = f(x,t+n)$ with $\psi_2(0) = \varphi(n)$. But $\|f(x,t+n+m) - f(x,t+n)\| < \delta$ so by the lemma $|\psi_1(t) - \psi_2(t)| < \epsilon$ for $0 \leq t \leq 1$. That is, $|\varphi(t+n+m) - \varphi(t+n)| < \epsilon$ for $0 \leq t \leq 1$. Since n is arbitrary we have $|\varphi(t+m) - \varphi(t)| < \epsilon$ for all t. Thus $T(\varphi,\epsilon) \supset T_1 \cap T_2$, so is relatively dense.

One way in which one gets the hypothesis about unique solutions to initial value problems is to have $f(x,t)$ Lipschitz in x uniformly in t. This carries over to the hull. In this form one can prove that φ is bounded from its almost periodicity on Z by a simple Gronwall inequality argument.

Question F. Is a theorem like Theorem 9.7 true for flows?

6. Separation by functionals. We would like to introduce a notion here that is in some sense a generalization of the idea of minimizing the norm. The idea of minimum norm separates out a special solution which proves to be almost periodic. We generalize by replacing the norm by some functional.

Definition 9.8. A functional λ mapping solutions onto the real numbers is said to be subvariant for the solutions of $x' = f(x,t)$

in K if λ is defined on solutions of all equations in the hull of f which lie in K and if $\lambda(T_\alpha\varphi) \leq \lambda(\varphi)$ for φ a solution.

The model for this notion is $\lambda(\varphi) = \|\varphi\|$.

<u>Lemma 9.9</u>. Suppose $f(x,t)$ is almost periodic in t uniformly for $x \in K$ and suppose there is a unique minimum of λ over solutions of $x' = f(x,t)$ in K, where λ is subvariant. If φ is the minimizing solution, then $\lambda(T_\alpha\varphi) = \lambda(\varphi)$ for all α such that $T_\alpha\varphi$ exists.

<u>Proof</u>: If $T_\alpha\varphi$ exists then for some subsequence $\beta \subset \alpha$ $\psi = T_\beta\varphi$ and $g(x,t) = T_\alpha f(x,t)$ exist so that ψ is a solution of $x' = g(x,t)$. There is a sequence γ so that $T_\gamma\psi$ and $T_\gamma g(x,t) = f(x,t)$ exist so that $T_\gamma\psi$ is a solution of $x' = f(x,t)$. Then $\lambda(T_\gamma\psi) \leq \lambda(\psi) = \lambda(T_\beta\varphi) = \lambda(T_\alpha\varphi) \leq \lambda(\varphi)$. By uniqueness of the minimum, $T_\gamma\psi = \varphi$ so equality holds throughout and $\lambda(T_\alpha\varphi) = \lambda(\varphi)$.

<u>Theorem 9.10</u>. Let $f(x,t)$ be almost periodic in t uniformly for $x \in K$. Suppose that a subvariant functional λ exists, for which the minimum over solutions is uniquely attained for each equation in the hull. If $x' = f(x,t)$ has a solution in K, then there is an almost periodic solution φ, with $\operatorname{mod}(\varphi) \subset \operatorname{mod}(f)$.

<u>Proof</u>: Let φ be the unique minimizing solution of $x' = f(x,t)$. If α' and β' are given, find common subsequences $\alpha \subset \alpha'$ and $\beta \subset \beta'$ so that $T_\alpha T_\beta f = T_{\alpha+\beta} f$ and $T_\alpha\varphi$, $T_\beta T_\alpha\varphi$, and $T_{\alpha+\beta}\varphi$ all exist uniformly on compact sets. Then $T_\beta T_\alpha\varphi$ and $T_{\alpha+\beta}\varphi$ are solutions of the same equation and $\lambda(T_\beta T_\alpha\varphi) = \lambda(\varphi) = \lambda(T_{\alpha+\beta}\varphi)$, by Lemma 9.9.

By Lemma 9.9, the minimizing value $\lambda(\varphi)$ is independent of the equation so $T_\beta T_\alpha \varphi$ and $T_{\alpha+\beta}\varphi$ both are minimizing solutions and $T_\beta T_\alpha \varphi = T_{\alpha+\beta}\varphi$, which proves the almost periodicity of φ. Furthermore, if $T_\alpha f = f$, then $\lambda(T_{\alpha'}\varphi) = \lambda(\varphi)$ for $\alpha' \subset \alpha$. Thus $T_{\alpha'}\varphi = \varphi$ and $\mathrm{mod}(\varphi) \subset \mathrm{mod}(f)$ by Theorem 4.5 v .

Note that without the hypothesis on the hull, namely just assuming uniqueness of minimizing solution for $x' = f(x,t)$, we can get $T_\alpha f = f$ implying $T_{\alpha'}\varphi = \varphi$ for some $\alpha' \subset \alpha$. This gives the almost automorphic property of φ.

<u>Question G.</u> Is Theorem 9.10 true by just assuming uniqueness of minimizing solution of $x' = f(x,t)$?.

An even more intriguing question is also available.

<u>Question H.</u> Can an almost periodic equation have an almost automorphic solution that is not almost periodic?

Note that Favard's Theorem, Theorem 6.3 is a corollary of Theorem 9.10. We will give some explicit applications of Theorem 9.10 in later chapters.

A generalization of Theorem 9.10 is also easy to state. We may for example have the range of λ be a partially ordered set, or we may assume that there are a finite number of subvariant functionals, so that the vector $(\lambda_1(\varphi),\ldots,\lambda_n(\varphi))$ is uniquely minimized in at least one coordinate. Another version might be that λ_2 is uniquely minimized along the solutions which minimize λ_1. We also give an application of this idea in a later chapter.

7. Notes. The notion of asymptotic almost periodic functions was introduced by Frechet with applications to ergodic theory. Frechet, Revue Sci. (Rev. Rose Illus.) 79(1941), 341-354, 407-417, and C. R. Acad. Sci. Paris 213(1941), 520-522, 607-609, proved most of the properties of such functions. Condition (5) in Theorem 9.3 is in Fink, Semi-separated conditions for almost periodic solutions, J. Differential Equations, 11(1972), 245-251.

Corollary 9.4 is found in Seifert, A condition for almost periodicity with some applications to functional-differential equations, J. Differential Equations 1(1965), 393-408. Theorem 9.5 is in Coppels book [5].

Meisters , On almost periodic solutions of a class of differential equations, Proc. Amer. Math. Soc. 10(1959), 113-119 first proved Theorem 9.7 under the condition that f is Lipschitz and this was replaced by uniqueness of initial value problems by Opial, Sur les solutions presque-periodiques d'une classe d'equations differentielles, Ann. Polon. Math. 9(1960/61), 157-181.

Chapter 10

Separated Solutions

1. Introduction. The motivation for what we are going to do in this chapter are the results of Favard, specifically Theorem 6.3. We will want to extend this to non-linear systems. The idea behind the hypothesis that $\inf_t |\varphi(t)| > 0$ for non-trivial solutions of the homogeneous system is this. One shows that the bounded solutions of the non-homogeneous system are a positive distance apart in the sense that $|\varphi_1(t) - \varphi_2(t)| \geq d > 0$ for all t for φ_1 and φ_2 solutions. It was the idea of Amerio that this separation condition was the crucial one. He showed that it led to almost periodicity of the solutions. We are not going to do Amerio's results here explicitly, but will derive analogous theorems. The idea is that stability of solutions in the future implies the separatedness of the solutions in the past. The equation $x' = x$ is a case in point.

2. The unique case. The only application of Amerio's theorems to specific cases that appear in the literature, are to the case when there is a unique solution in some compact set K. Since this version has a simple and direct proof we give it here.

Theorem 10.1. Suppose $f(x,t)$ is almost periodic in t uniformly for x in K, K compact in E^n. If each equation $x' = g(x,t)$, $g \in H(f)$ has a unique solution on R with values in K, then these solutions are almost periodic with module contained in $\mathrm{mod}(f)$.

Proof: This should be very familiar by now. Let φ be the solution of $x' = f(x,t)$ in K. If α' and β' are given, take $\alpha \subset \alpha'$, $\beta \subset \beta'$ common subsequences so that $T_{\alpha+\beta}f = T_\alpha T_\beta$, and so that $T_\alpha T_\beta \varphi$ and $T_{\alpha+\beta}\varphi$ exist uniformly on compact sets. Then $T_\alpha T_\beta \varphi$ and $T_{\alpha+\beta}\varphi$ are solutions in K of the same equation in H(f). Thus $T_\alpha T_\beta \varphi = T_{\alpha+\beta}\varphi$ and φ is almost periodic. Since $T_\alpha f = f$ implies $T_\alpha \varphi = \varphi$, we have module containment.

Corollary 10.2. Let f be as in Theorem 10.1 and suppose the hypothesis on solutions replaces R by R^+, then we have the conclusion that there is an almost periodic solution on R which agrees with the given solution on R^+.

Proof: By applying Theorem 6.2 we have the existence of a solution on R in K. But this solution must be the original one on R^+. Now apply Theorem 10.1. The Corollary 10.2 has weaker hypothesis but those of the theorem follow more naturally from stability conditions, see Chapter 11 below.

3. Semi-separated solutions. We will say that a solution φ in K, of

$$x' = f(x,t) \tag{10.1}$$

is semi-separated in K if there is a number $d(\varphi) > 0$ so that if ψ is any other solution in K of (10.1), then $|\varphi(t) - \psi(t)| \geq d(\varphi)$ on R^+.

If the last inequality were to hold on R, we say that φ is separated.

Lemma 10.3. Suppose $f(x,t)$ is almost periodic in t uniformly for x in K, K compact. If every equation in $H(f)$ has only semi-separated solutions in K, then each such equation has only a finite number of solutions in K. Consequently, this number is the same for every equation in $H(f)$ and the constant $d(\varphi)$ may be picked independent of solution or equation.

Proof: Fix an equation in the hull. It has only a finite number of solutions in K. Since the set of solutions are uniformly bounded and equi-uniformly continuous, an infinite number of solutions would contain a sequence that converges uniformly on compact sets to some solution in K. That limit solution could not be semi-separated. Now that the number is finite, we may select the constant $d(\varphi)$ to be dependent only on the equation so we switch the notation to $d(g)$, $g \in H(f)$. If g_1 and g_2 are in $H(f)$, there is α' so that $T_{\alpha'}(id) = +\infty$ and $T_{\alpha'}g_1 = g_2$. If $\varphi_1,\ldots,\varphi_n$ are the solutions of $x' = g_1(x,t)$ in K, we may find $\alpha \subset \alpha'$ so that $\psi_i = T_\alpha \varphi_i$, $i = 1,\ldots,n$, exist and are solutions of $x' = g_2(x,t)$ in K. Since $|\varphi_i(t) - \varphi_j(t)| \geq d(g_1)$ if $i \neq j$ and $t \in R^+$, we have the same conclusion $|\psi_i(t) - \psi_j(t)| \geq d(g_1)$ if $i \neq j$, $t \in R^+$. Thus the equation $x' = g_2(x,t)$ has at least n solutions in K. By symmetry $x' = g_1(x,t)$ has at least as many as $x' = g_2(x,t)$, so the number is the same. Knowing this, we see that the ψ_i exhaust the solutions of $x' = g_2(x,t)$ in K and so $d(g_2) \geq d(g_1)$. Again by symmetry, we get equality.

Theorem 10.4. Let $f(x,t)$ be almost periodic in t uniformly for $x \in K$, K compact. Suppose each equation in $H(f)$, $x' = g(x,t)$ say, has only semi-separated solutions in K. If there is a solution of (10.1) in K, then all solutions of (10.1) in K are a.a.p. so that there are almost periodic solutions.

Proof: Let φ be a solution of (10.1) in K. Let d be the separation constant. The claim that φ is a.a.p. will be proved by showing that (5) of Theorem 9.3 holds with d replaced by $d/2$. To see this, let α' be a sequence such that $T_{\alpha'}(id) = +\infty$. Let $\alpha \subset \alpha'$ such that $T_\alpha f$ and $T_\alpha \varphi$ exist uniformly on compact sets. Take $d(\alpha) = d/2$. If δ, β, and γ are sequences so that $\delta > 0$, $\beta \subset \alpha$, $\gamma \subset \alpha$ and $T_{\delta+\beta}\varphi = \psi_1$ and $T_{\delta+\gamma}\varphi = \psi_2$ exist, we may assume by taking subsequences that $T_{\delta+\beta}f = T_\delta T_\beta f = T_\delta T_\alpha f$ and $T_{\delta+\gamma}f = T_\delta T_\gamma f = T_\delta T_\alpha f$. Thus ψ_1 and ψ_2 are solutions of the same equation. Either $|\psi_1(t) - \psi_2(t)| \geq d$ or $\psi_1 \equiv \psi_2$ on R^+, since the separation constant d is independent of solution or equation. Thus by (5) φ is a.a.p. By Theorem 9.2, the almost periodic part of φ is a solution of (10.1).

Several questions arise naturally about the necessity of the hull hypothesis.

Question I: Is Theorem 10.4 true if only $x' = f(x,t)$ is required to have semi-separated solutions? The question is even of more interest in the case of Theorem 10.1.

The answers to the above are probably negative, but it

would be nice to have a specific example. Note also that the module containment is missing from Theorem 10.4 but is in Theorem 10.1. The reason is that it does not follow.

<u>Example 10.5</u>. Let $A(t) = \begin{pmatrix} \cos t & \sin t \\ -\sin t & \cos t \end{pmatrix}$, and $\xi_1' = \xi_1$, $\xi_2' = \xi_2(1-\xi_2^2)$. Write this latter system as $\xi' = f(\xi)$ where $\xi = (\xi_1,\xi_2)^T$, and consider $x = A(t)\xi$. Then $x' = f(x,t) = A'A^Tx + Af(A^Tx,t)$. Using the facts that f is odd, $A(t+\pi) = -A(t)$, $A'(t+\pi) = -A'(t)$ and $A^T(t+\pi) = -A^T(t)$ we have that $f(x,t+\pi) = f(x,t)$. Introduce a second function

$$g(x_1,x_2,x_3) = \begin{cases} x_3(1-2(x_1^2+x_2^2)) & \text{if } x_1^2+x_2^2 > \frac{1}{2} \\ 1-2(x_1^2+x_2^2) & \text{if } x_1^2+x_2^2 \le \frac{1}{2}. \end{cases} \tag{10.2}$$

Writing $x = (x_1,x_2)^T$ introduce the three dimensional system

$$\begin{cases} x' = f(x,t) \\ x_3' = g(x_1,x_2,x_3) \end{cases} \tag{10.2}$$

This system has the two periodic solutions $\varphi_1 = (-\sin t,\cos t,0)^T$ and $\varphi_2 = (\sin t,-\cos t,0)^T$ each with period 2π. They are separated, since $|\varphi_1(t)-\varphi_2(t)| \equiv \sqrt{2}$ for all t, but $\operatorname{mod}(\varphi_i) \not\subset \operatorname{mod}(f)$, the former being the integers Z, and the latter $2Z$. Let $K = \{(x_1,x_2,x_3)^T : x_3 = 0 \ x_1^2+x_2^2 = 1\}$. We need to show that there are no other solutions in K. Note that $x_1^2+x_2^2 = \xi_1^2+\xi_2^2$ so that $x_1^2+x_2^2 = 1$ implies that $\xi_1^2+\xi_2^2 = 1$. In particular, ξ, is bounded. For our purposes here we can think of R^- instead of R^+ on which we want semiseparatedness. Since $\xi_1(t) = \xi_1(0)e^{-t}$ is to be bounded on R^- we must have $\xi_1(t) \equiv 0$, and thus $\xi_2(t) \equiv 1$ or $\xi_2(t) \equiv -1$. These correspond to φ_1 and φ_2 above.

The above example is unnecessarily complicated. We will, however, use it later to illustrate something else. It turns out that the φ_i are uniformly asymptotically stable.

4. A generalization. The results of the preceding section have at least two weaknesses. One is the need for the hypothesis on all equations in the hull, and the other is the need for the solutions in question to be separated from all the other solutions in K. We intend to offer partial remedies here. The idea is that we want to find a subclass of solutions for which the arguments of Lemma 10.3 and Theorem 10.4 work. This requires that the subclass be closed under the operation of translation, namely T_α, and that the solutions in this subclass be semi-separated. The working model we have, is that the subclass is the set of uniformly stable solutions. We will show that this model fits into the general framework in the next chapter.

To set the problem we will consider the equation (10.1) with the function satisfying $H_1 : f(x,t)$ is almost periodic in t uniformly for $x \in K$ where K is a fixed compact set. Furthermore, if a solution is mentioned, it has values in K.

As we have noted before, stability in the future implies semi-separtedness in the past, so we will replace R^+ by R^- here to conform with that model. Furthermore, a.a.p. will also be considered on R^-, that is, $q(t) \to 0$ as $t \to -\infty$. The reader should have no difficulty reformulating the previous theorems to apply to this situation.

Definition 10.6. A property P is said to be inherited if when φ is a solution of $x' = f(x,t)$ having property P with respect to the solutions if $x' = f(x,t)$ and $g(x,t) = T_\alpha f(x,t)$, $T_\alpha \varphi = \psi$ exist, then ψ has property P with respect to the solutions of $x' = g(x,t)$.

Definition 10.7. A property P of a solution φ of (10.1) is a semi-separating property if for any other solution ψ of (10.1) with property P, there is a number $d(\varphi,\psi) > 0$ so that $|\varphi(t) - \psi(t)| \geq d(\varphi,\psi)$ on R^-.

Note that the inherited property is to ensure that the solutions with property P are closed under T_α. The semi-separating property is self-explanatory. One other thing to note is that d may depend on both solutions whereas in Lemma 10.3 d could only depend on one.

The major point of introducing these notions is that we can prove the analogues of Lemma 10.3 and Theorem 10.4

Lemma 10.8. Suppose that property P is inherited and is semi-separating. If (10.1) has only a finite number of solutions in K with property P, then every equation in the hull has the same number of solutions in K with property P and the separation constant may be picked independent of solution and equation.

Proof: Say $x' = f(x,t)$ has n solutions in K with property P. Then the separation constant can be picked independent of these solutions, say $d(f)$. If φ and ψ are solutions of $x' = f(x,t)$ with property P, $T_{\alpha'}(id) = -\infty$, and $T_{\alpha'}f = g$, then for $\alpha \subset \alpha'$ $T_\alpha\varphi$ and $T_\alpha\psi$ are solutions of $x' = g(x,t)$ in K with property P. Furthermore, $|T_\alpha\varphi(t) - T_\alpha\psi(t)| \geq d(f)$. In particular $T_\alpha\varphi \neq T_\alpha\psi$. Thus if $\varphi_1,\ldots,\varphi_n$ are the solutions of $x' = f(x,t)$ with property P, then $T_\alpha\varphi_1,\ldots,T_\alpha\varphi_n$ are distinct semi-separated solutions of $x' = g(x,t)$ with property P. Thus $x' = g(x,t)$ has at least n such solutions. If it had more, take $(n+1)$ of them and get $(n+1)$ for $x' = f(x,t)$ by the above argument. Thus it has exactly n. Now the proof that the separation constant is independent of equation is the same as before.

Theorem 10.9. If P is a property that is inherited and semi-separating and (10.1) has only a finite number of solutions in K with property P, then these solutions are almost periodic.

Proof: Given the lemma, the proof is exact replica of the proof of Theorem 10.4 with the slight exception of replacing R^+ by R^-.

This generalization is of interest because the hypotheses are solely on the equation $x' = f(x,t)$, with the hull hypothesis being replaced by the property of inheritance of P, and in the permission of the separation constant to depend on both solutions. Examples are given in the next chapter. Of course Theorem 10.1

is a Corollary of these results.

5. Notes. The results of Amerio are in the landmark paper Amerio, Soluzioni quasi-periodiche, o limitate, di sistemi differenziali non lineari quasi-periodici, o limitate, Ann. Mat. Pura Appl. 34(1955), 97-119.

The results in the form of this chapter are in Fink, Semi-separated conditions for almost periodic solutions, J. Differential Equations, 11(1972), 245-251. Note that the semi-separated property is reminscent of the distal property in topological dynamics.

Chapter 11

Stable Solutions.

1. Introduction. We are to give an exposition of the relationship between almost periodicity and stability. The subject is an old one and we give one of the earliest examples first. Then we will show how stability implies almost periodicity and how we can use stability to perturb equations. We will not go deeply into the use of Lypanov's method to get stability since excellent monographs on the subject already exist.

2. Strong stability. The idea to be exposed in this section is the condition that "point compactness implies orbit compactness" for a solution implies its almost periodicity. In order to do this clearly we first derive a new equivalence for almost periodicity.

Definition 11.1. A subset S of R is said to be a Δ-m set if $S = -S$ and for every set of $(m+1)$ real numbers $\{t_1,\ldots,t_{m+1}\}$ one can extract $i \neq j$ such that $t_i - t_j \in S$.

Note that we do not require that the numbers t_i be distinct so that $0 \in S$ is possible. Clearly such sets exist. In fact if $f \in AP(E^n)$ then $T(f,\epsilon)$ satisfies this property.

Lemma 11.2. If $f \in AP(E^n)$ then $T(f,\epsilon)$ is a Δ-m set for some positive integer m.

Proof: We use the total boundedness of $F = \{f_a \colon f_a(t) = f(a+t)\}$. If ϵ is given find $a_1,\ldots,a_m$ so that every element of F is within $\epsilon/2$ of one of the function f_{a_i}, $i = 1,\ldots,m$. The claim is that $T(f,\epsilon)$ is a Δ-m set, for if $t_1,\ldots,t_{m+1}$ are given then two of the functions $f_{t_1},\ldots,f_{t_{m+1}}$ are within $\epsilon/2$ of the same f_{a_j} so by the triangle inequality $\|f_{t_i} - f_{t_j}\| < \epsilon$ for some $i \neq j$. This means $|f(t+t_i) - f(t+t_j)| < \epsilon$ for all t and equivalently $|f(s+t_i-t_j) - f(s)| < \epsilon$ for all s. That is, $t_i - t_j \in T(f,\epsilon)$ as was to be shown.

The interesting thing about the concept of Δ-m set is that the correct formulation of Lemma 11.2 has a true converse.

Theorem 11.3. Let f be continuous. Then $f \in AP(E^n)$ if and only if for every $\epsilon > 0$ $T(f,\epsilon)$ is a Δ-m set for some m.

Proof: Half of the statement is Lemma 11.2. To prove the other half we need to prove that Δ-m sets are relatively dense. In order to do this we use induction on m. Note $T(f,\epsilon) = -T(f,\epsilon)$ in any case.

The first claim is that the only Δ-1 set is R. In fact, consider $\{t,0\}$ for $t \in R$. If S is a Δ-1 set then either t or $-t \in S$. But since S is symmetric, both are, thus $t \in S$.

We now assume that all Δ-m sets are relatively dense and try to show that all Δ-(m+1) sets are relatively dense. Let S be a Δ-(m+1) set. If S is also a Δ-m set we are done by the hypothesis. So assume S is not a Δ-m set. There are numbers $t_1,\ldots,t_{m+1}$ so that $t_i - t_j \notin S$ for $i \neq j$. If S is not relatively dense, there is an interval I whose length is greater than $2 \max|t_i|$

with $S \cap I = \phi$. It is thus possible to find a number x so that $u_i = x + t_i \in I$ for $i = 1,\ldots,m+1$. Now consider the $(m+2)$ real numbers $\{0,u_1,\ldots,u_{m+1}\}$. Then $u_i - u_j = t_i - t_j \notin S$ by hypothesis, and $u_i - 0 \in I$ so not in S. Since S is symmetric, $0 - u_i \notin S$. This contradicts the hypothesis that S is a Δ-$(m+1)$ set.

Definition 11.4. A motion $x(t)$ is said to be strongly stable if for every $\epsilon > 0$ there is a $\delta > 0$ so that if $|x(t_1) - x(t_2)| < \delta$ then $|x(t+t_1) - x(t+t_2)| < \epsilon$ for all t.

What one has in mind is that x is a solution to an autonomous differential equation $x' = f(x)$ in which case $x(t+t_1)$ and $x(t+t_2)$ are both solutions that are close at 0, so strong stability requires that they consequently stay close, both in the future and in the past.

Theorem 11.5. If x is a continuous bounded strongly stable motion in E^n, then $x \in AP(E^n)$.

Proof: Since x is bounded in E^n, if $\epsilon > 0$ is given we find numbers $a_1,\ldots,a_m$ so that every vector in $\{x(t): t \in R\}$ is within $\delta/2$ of one of the vectors $x(a_1),\ldots,x(a_m)$, where $\delta = \delta(\epsilon)$ comes from the strong stability (11.4). This is possible since bounded sets in E^n are totally bounded. Now consider $T(x,\epsilon)$. The claim is, that it is a Δ-m set. For if $t_1,\ldots,t_{m+1}$ are given then two of them say t_i, t_j satisfy $|x(t_i) - x(a_{n_0})| < \delta/2$ and $|x(t_j) - x(a_{n_0})| < \delta/2$. Consequently, $|x(t_i) - x(t_j)| < \delta$ and

$|x(t+t_i) - x(t+t_j)| < \epsilon$ for all t. That is $t_i - t_j \in T(x,\epsilon)$ so x is in $AP(E^n)$.

One can see that the points $\{x(t)\}$ indexed by $t \in R$ have compact closure, and the strong stability gives the same property to the set $\{x(s+t)\}$ in $C(R)$, indexed by $t \in R$.

3. Uniform Stability. The notion of strong stability is not useful in that many systems of interest do not satisfy this kind of property. It is more natural to consider stability only in the future. We will not attempt a comprehensive catalogue of stability notions but talk only about those which we find useful in the context of a.p. differential equations.

Definition 11.6. A solution φ of $x' = f(x,t)$ is uniformly stable on $[t_1,\infty)$ if for every $\epsilon > 0$, there is a $\delta > 0$ so that $t_0 \in [t_1,\infty)$ and $|\varphi(t_0) - \psi(t_0)| < \delta$ for ψ a solution of $x' = f(x,t)$ imply that $|\varphi(t) - \psi(t)| < \epsilon$ on $[t_0,\infty)$.

Definition 11.7. A solution φ of $x' = f(x,t)$ is uniformly asymptotically stable on $[t_1,\infty)$ if it is uniformly stable and if there is an $r > 0$ so that for every $\epsilon > 0$, there is a $T(\epsilon) > 0$ so that if, $t_0 \geq t_1$ and $|x_0 - \varphi(t_0)| < r$ then $|x(t;t_0;x_0) - \varphi(t)| < \epsilon$ on $t \geq t_0 + T(\epsilon)$. Here $x(t;t_0;x_0)$ is a solution of $x' = f(x,t)$ with $x(t_0;t_0;x_0) = x_0$.

Uniform stability is continuity with respect to initial conditions on $[t_0,\infty)$ where the δ is independent of t_0, while

$\lim_{t\to\infty} [\varphi(t) - x(t;t_0;x_0] = 0$ is added in uniform asymptotic stability in a very precise way. If T is allowed to depend on t_0, then one has asymptotic stability. We will introduce several other notions of stability later, but we first want to show how these two types lend themselves to proofs of existence of almost periodic solutions. Stability on R is interpreted as $t_1 = -\infty$.

The major themes of this section are that both of the above types of stability are inherited and semi-separating. In order to obviate the necessity of repeating a long hypothesis every time we introduce the standard hypothesis.

Standard Hypothesis. The equation

$$x' = f(x,t) \tag{11.1}$$

is said to satisfy the standard hypothesis if

(i) $f(x,t)$ is almost periodic in t uniformly for $x \in K$ where K is a fixed compact set in E^n, and

(ii) every equation in the hull of (11.1) has unique solutions to initial value problems in K.

We also let the notation $x_G(t;t_0;x_0)$ mean that this is a solution of $x' = G(x,t)$, $x(t_0) = x_0$.

Lemma 11.8. Let (11.1) satisfy the standard hypothesis. If φ is a uniformly stable solution of (11.1) on $[t_1,\infty)$, and if $T_\alpha(\mathrm{id}) = +\infty$, $T_\alpha f = g$, and $T_\alpha \varphi = \psi$ exist uniformly on compact sets, then ψ is a uniformly stable solution of $x' = g(x,t)$ on R.

Proof: If $\epsilon > 0$ is given, let δ be chosen as in definition 11.6. Now let $t_0 \in R$, and $x_0 \in K$ such that $|x_0 - \psi(t_0)| < \delta/2$. Let N be so large that if $n \geq N$, then $t_0 \geq t_1 - \alpha_n$ and $|x_0 - \varphi(t_0 - \alpha_n)| < \delta$. Then

$$|x_f(t+\alpha_n; t_0+\alpha_n; x_0) - \varphi(t+\alpha_n)| < \epsilon \tag{11.2}$$

if $t \geq t_0$. Fix $t \geq t_0$. Then $y_n(t) = x_f(t+\alpha_n; t_0+\alpha_n; x_0)$ converges to $y(t)$, a solution of $x' = T_\alpha f(x,t) = g(x,t)$. Since $y_n(t_0) = x_0$ for all n, $y(t) = x_g(t; t_0; x_0)$. Thus taking limits in (11.2) we get $|x_g(t; t_0; x_0) - \psi(t)| \leq \epsilon$. Since $t \geq t_0$ is arbitrary, we have that ψ is uniformly stable on R (t_0 was also arbitrary).

Corollary 11.9. Under the standard hypothesis, uniform stability is an inherited property.

Corollary 11.10. If (11.1) has a uniform stable solution on $[t_1, \infty)$ and the standard hypothesis is satisfied, then (11.1) has a uniformly stable solution on R.

Proof: We just apply Lemma 1.8 twice, since there is a sequence β so that $T_\beta(\mathrm{id}) = +\infty$ and $T_\beta g = f$.

Lemma 11.11. Suppose (11.1) satisfies the standard hypothesis and has a uniformly asymptotically stable solution φ on $[t_1, \infty)$. If $T_\alpha f = g$ and $T_\alpha \varphi = \psi$ with $T_\alpha(\mathrm{id}) = +\infty$, then ψ is a uniformly asymptotically stable solution of $x' = g(x,t)$ on R.

<u>Proof</u>: This proof parallels that of Lemma 11.8, indeed, the uniform stability of ψ on R follows from that lemma so we need to show the asymptotic relation. We refer to the proof of Lemma 11.8 and its notations. If $r > 0$ is given by definition 11.7, we take x_0 so that $|x_0 - \psi(t_0)| < r - r_0$ for some $r_0 \in (0,r)$. For large n, $|x_0 - \varphi(t_0 + \alpha_n)| < r$ so that $|x_f(t+\alpha_n; t_0 + \alpha_n; x_0) - \varphi(t+\alpha_n)| < \epsilon/3$ if $t+\alpha_n \geq t_0 + \alpha_n + T(\epsilon/3)$. Then for t fixed with $t \geq t_0 + T(\epsilon/3)$ we take limits as in Lemma 11.8 to get $|x_g(t; t_0; x_0) - \psi(t)| \leq \epsilon/3$ if $t \geq t_0 + T(\epsilon/3)$, as was to be shown.

We again have the corollaries corresponding to 11.9 and 11.10.

<u>Corollary 11.12</u>. Under the standard hypothesis, uniform asymptotic stability is an inherited property.

<u>Corollary 11.13</u>. If (11.1) has a uniform asymptotically stable solution on $[t_1, \infty)$, then it has a solution that is uniformly asymptotically stable solution on R.

In order to apply the results of the preceding chapter we need to know that the above types of stability are semi-separating. This turns out to be very easy.

<u>Lemma 11.14</u>. Under the standard hypothesis, uniform stability and uniform asymptotic stability on R are semi-separating.

<u>Proof</u>: If φ is a uniformly stable solution of (11.1) on R, let ψ be another solution in K. Then for $\epsilon = |\psi(0) - \varphi(0)| > 0$ we have $|\psi(t) - \varphi(t)| > \delta$ on R^- with δ from definition 11.6.

Indeed, if for some $t \in R^-$ $|\psi(t) - \varphi(t)| < \delta$ then $|\psi(0) - \varphi(0)| < \epsilon$. Since uniformly asymptotic stability includes uniform stability, the other half of the Lemma is also true.

Theorem 11.15. Let (11.1) satisfy the standard hypothesis. If (11.1) has only a finite number of solutions that are uniformly stable (or uniformly asymptotically stable) on $[t_1, \infty)$ then it has an almost periodic solution which is uniformly stable (uniformly asymptotically stable) on R.

Proof: We first apply Corollary 11.10 or 11.13 to get a finite number of uniformly stable (or uniformly asymptotically stable) solutions on R. Then the property P of being uniformly stable (or uniformly asymptotically stable) is inherited and semi-separating on R^- so we apply Theorem 10.9 to get the almost periodicity of all of these solutions.

It is possible to have an infinite number of uniformly stable solutions in a compact set.

Example 11.16. The equation $x' = x^2 \sin 1/x$, $x \neq 0$, $x' = 0$ if $x = 0$, has infinitely many uniformly stable solutions. The solution $x(t) \equiv (n\pi)^{-1}$ is uniformly stable if n is even.

The solutions $(n\pi)^{-1}$ are also uniformly asymptotically stable so it is an interesting question whether this is necessary. Note that the zero solution is not asymptotically stable.

Question J. Is it possible to have an infinite number of uniformly stable solutions in a compact set, none of which are uniformly asymptotically stable? Yes, see 216.

Several other questions arise in the context of these results. Is the standard hypothesis needed? In particular, does unique solution of initial value problems carry over to the equations in the hull? Or, is uniform stability inherited without unique solutions of initial value problems? The answer to both questions is, no. We give examples.

Example 11.17. Consider the scalar equation $x' = g(x,t)$ with g periodic of period π in t where

$$g(x,t) = \begin{cases} 0 & -\pi/2 \leq t \leq 0 & |x| < \infty \\ 0 & 0 \leq t \leq \pi/2 & x \leq 0 \\ 2x \cot t & 0 \leq t \leq \pi/2 & 0 \leq x \leq \sin^2 t \\ 2 \sin t \cos t & 0 \leq t \leq \pi/2 & \sin^2 t \leq x \end{cases}$$

If $(x_0,t_0) \neq (0,k\pi)$ there are unique solutions to the problem through (x_0,t_0) but $x(t,c) = c \sin^2 t$, $0 \leq c \leq 1$, $0 \leq t \leq \pi/2$ are solutions through $(0,0)$ and similar ones through $(0,k\pi)$. Now suppose $p(t)$ is continuous and negative at $k\pi$. Then $x' = g(x,t) + p(t)$ has unique solutions everywhere since Lipschitz conditions hold everywhere except at $(0,k\pi)$ but at $t = k\pi$, $x' = p(t)$ for $t \geq k\pi$ and $t - k\pi$ small. Thus these solutions are also unique. To construct the example we consider $x' = g(x,t) + q(t)$ with q continuous and period 2, such that $q(t) < 0$ on $(0,2)$ and $q(t) = -\sin^3 t$ on $0 \leq t \leq \pi/2$. The equation is then almost periodic and has two solutions through $(0,0)$ namely $x_1(t) = \cos t \sin^2 t > 0$

and $x_2(t) = -\int_0^t \sin^3 u\, du = -2/3 + \cos t - 1/3 \cos^3 t < 0$ on $(0,\pi/2)$. At all other points $(0,k\pi)$ $q < 0$ so there are unique solutions there. There are equations in the hull that have unique solutions everywhere. For example, let k_n be a sequence so that $k_n\pi \to 1 \bmod(2)$. Then $g(x,t+kn) + q(t+kn) \to g(x,t) + q(t+1) = f(x,t)$. Now $x' = f(x,t)$ has unique solutions of initial value problems since $q(k\pi+1) < 0$ for all k.

<u>Example 11.18</u>. We first construct a scalar almost periodic function $a(t)$ which is positive everywhere, but has a function in the hull which is zero on an interval. To do this, let $b_0(t) = 1$ and for $n > 1$, $b_n(t) = 0$ on $0 \le t \le 2^n$; $b_n(t) = -2^{-n}$ for $t \in (2^n+\epsilon_n, 2^{n+1}-\epsilon_n)$; extend to be odd and periodic with period 2^{n+2} except that it is to be smoothed to make it continuous with $\|b_n\| = 2^{-n}$, and $b(t) \le 0$. Then $a(t) = \sum_{n=0}^{\infty} b_n(t)$ is the required function. It is easy to see that $a > 0$ and that it is almost periodic. Note that $a(t+2^{n-1}) = 2^{-n}$ for $0 \le t \le 1$ and $n \ge 2$. Thus if $\alpha'_n = 2^{-n}$ we find $\alpha \subset \alpha'$ so $T_\alpha a = b$, then $b(t) = 0$ on $[0,1]$. Now consider

$$f(x) = \begin{cases} 4-x & 2 \le x \le 4 \\ 2\sqrt{|nx-1|} & \frac{2}{2n+1} \le x \le \frac{2}{2n-1} \quad n = 1,2,\ldots; \\ 0 & x = 0 \quad \text{or} \quad x \ge 4 \\ \text{odd.} & \end{cases}$$

and the differential equation

$$x' = \begin{cases} f(x) - c\,a(t)\sqrt{x}\,, & x \geq 0\,; \\ \text{odd in } x. \end{cases}$$

Suppose $c \geq 2\sqrt{2}$. Since $a(t) > 0$, we have $x'(\frac{1}{n}) = -c\,a(t)\sqrt{\frac{1}{n}} < 0$ and $x'(-\frac{1}{n}) > 0$. Thus no solutions escape $[-\frac{1}{n},\frac{1}{n}]$ as t increases. Thus the zero solution is uniformly stable. But if we consider α as above then the equation

$$x' = \begin{cases} f(x) - c\,b(t)\sqrt{x}\,, & x \geq 0; \\ \text{odd in } x. \end{cases} \tag{11.3}$$

has zero as a solution but it is not uniformly stable. To see this we will construct a solution to this equation which passes through $(0,0)$, is defined on $t \geq 0$, and is not identically zero. Note that on $[0,1]$ we are looking at the equation $x' = f(x)$. We first look at a solution of this equation which is defined piecewise. One verifies that if $x(t_0) = \frac{1}{n}$ and $x(t_2) = \frac{2}{2n-1}$ for $t_0 < t_2$ and $x' = f(x)$, then $t_2 - t_0 = \frac{1}{n}\frac{1}{\sqrt{2n-1}}$ since $x(t) = \frac{1}{n} + n(t_0 - t)^2$ on that interval. Similarly, if $t_1 < t_0$ and $x(t_1) = \frac{2}{2n+1}$, then $t_0 - t_1 = \frac{1}{n\sqrt{2n+1}}$. Thus if x is a solution of $x' = f(x)$ so that $x(t_n) = \frac{2}{2n-1}$, then $t_{n+1} = t_n - \frac{1}{n}\left(\frac{1}{\sqrt{2n-1}} + \frac{1}{\sqrt{2n+1}}\right)$. Thus

$$t_n = t_1 - \sum_{k=1}^{n-1} \frac{1}{k}\left(\frac{1}{\sqrt{2k-1}} + \frac{1}{\sqrt{2k+1}}\right). \quad \text{Let} \quad s = \sum_{k=1}^{\infty} \frac{1}{k}\left(\frac{1}{\sqrt{2k-1}} + \frac{1}{\sqrt{2k+1}}\right)$$

and $t_1 = s$. Then $t_n \to 0$, $x'(t_n) = 2\sqrt{2}\ (2n-1)^{-1/2} \to 0$ and $x(t_n) \to 0$. The odd extension of x defined to be 0 at 0 is now a solution of $x' = f(x)$, $x(0) = 0$ on $[-s,s]$. This function is a solution of (11.3) on $[0,1)$. It can be extended to be 0 to the left and can be extended to the right. Since $x' \leq 0$ if $x \geq 4$ in (11.3) and $x' \geq 0$ if $x \leq -4$, solutions can be continued to all of R. Hence the 0 solution is not uniformly stable.

Now it can be seen that by the methods developed so far we are unable to deduce the almost periodicity of uniformly stable solutions without the standard hypothesis. George Sell has indicated to that for scalar equations, one can prove that if (11.1) has a uniformly stable solution then it has an almost periodic solution provided (11.1) has unique solutions to initial value problems.

4. Stability of Linear Systems. If (11.1) is the linear equation

$$x' = A(t)x \tag{11.4}$$

then uniform stability of the 0 solution is equivalent to having an exponential dichotomy with $P = I$ and $\sigma_1 = 0$, while uniform asymptotic stability requires $\sigma_1 > 0$. If either one of these is satisfied then a bounded solution of

$$x' = A(t)x + f(t) \tag{11.5}$$

will have the same stability property, see Coppel [5].

Slightly more can be said in special cases of (11.4). For example, if A is constant, then stability and uniform stability are equivalent because of the translation invariance of the equation. Now the same is true of periodic systems. The reason for this is an application of Floquet Theory. If A is periodic then (11.4) can be transferred to

$$y' = B\,y \tag{11.6}$$

where B is a constant via $y = Px$ with P and P^{-1} bounded on R. This boundedness of P gives equivalences of the stability notions in (11.4) and (11.6). Thus (11.6) has the property that stability and uniform stability are equivalent which transfers to the periodic (11.4).

This points out the usefulness of a Floquet Theory, one we would like for almost periodic systems. However, we are now in a position to verify that no Floquet Theory holds for all almost periodic systems. In particular we describe an equation which has 0 as an asymptotically stable solution but which is not uniformly asymptotically stable.

<u>Example 11.19</u>. We construct a scalar almost periodic function f such that

(a) $\lim_{T\to\infty} \int_0^T f(s)\,ds = +\infty$ and

(b) there are sequences t, S with $T_S(id) = +\infty$ such that $\int_{t_n}^{t_n+S_n} f(s)\,ds \to -\infty$.

The function f is to be defined as $\sum_{n=1}^{\infty} g_n(t)$ where g_n is periodic with period 2^n and g_n is piecewise linear, 0 on $[0,2^{n-1}]$ and with a trapezoidal graph on $[2^{n-1},2^n]$. The exact determination is to be made in a moment. In order to set the problem, suppose $m_n = \frac{1}{2^n}\int_0^{2^n} g_n(s)ds$ and $a_n = \sum_{j=1}^{n} m_j$. Then $\int_0^{2^n} f(s)ds = 2^n a_n$ since $g_m \equiv 0$ on $[0,2^n]$ if $m \geq n$; and if $m < n$, g_m has a period that divides 2^n so that $m_m = \frac{1}{2^n}\int_0^{2^n} g_m(s)ds$. Also $\int_{2^n}^{2^{n+1}} f(s)ds = 2^{n+1}a_{n+1} - 2^n a_n = 2^n[2a_{n+1} - a_n]$. It is thus sufficient to pick numbers a_n (from which m_n can be recovered) so that

(i) $2^n a_n \to +\infty$; and along some subsequence

(ii) $\lim_n 2^n(2a_{n+1} - a_n) = -\infty$.

First we show that if any sequence a_n is given, $a_0 = 0$, and $A > 2$. Then there exist the functions g_n so that $\|g_n\| \leq A|m_n|$, where m_n is mean value of g_n and g_n is as described above. Let $g_n \equiv 0$ if $m_n = 0$, otherwise let $g_n \equiv 0$ on $[0,2^{n-1}]$; $\operatorname{sgn}(m_n)\mu_n$ $(\mu_n > 0)$ on $[2^{n-1}+\epsilon_n, 2^n - \epsilon_n]$ and extended to be continuous by being piecewise linear. Then a straightforward computation shows that $m_n = \mu_n[\frac{1}{2} - \frac{\epsilon_n}{2^n}]$ so we need to show that ϵ_n can be picked so that $[\frac{1}{2} - \frac{\epsilon_n}{2^n}]^{-1} < A$. One choice is $\epsilon_n = \frac{\delta}{2+\delta}$ where $A = 2+\delta$. Next observe that f will be almost periodic if $\sum|m_n| < \infty$, since $\sum \|g_n\| \leq A \sum |m_n|$. Thus we need to define a sequence a_n so that this is satisfied as well as (i) and (ii). Define $a_n = a_{n-1}$ if n

is even and $a_n = B\,a_{n-1}$ if n is odd with $a_0 = 0$ and $a_1 = 1$, and $1/4 < B < 1/2$. The sequence looks like $0,1,1,B,B,B^2,B^2,\ldots$. It is easy to see that $a_n \geq B^{n/2}$ so that $2^n a_n \geq (2\sqrt{B})^n \to +\infty$ since $2\sqrt{B} > 1$. Thus (i) is satisfied. Next, if n is even, then

$2^n[2a_{n+1} - a_n] = 2^n[2B-1]a_n \to -\infty$ since $2B-1 < 0$. Thus (ii) is satisfied. Finally, we need $\sum |m_n| < \infty$. Note that if n is even then $m_n = 0$ so we restrict ourselves to odd n. Then

$$m_n = a_n - a_{n-1} = B\,a_{n-1} - a_{n-2} = B\,a_{n-2} - B\,a_{n-3} = B(a_{n-2} - a_{n-3}) = B\,m_{n-2}$$

so $|m_n| < 1/2\,|m_{n-2}|$ and by the ratio test, the series converges. To complete the construction of f such that (a) and (b) are satisfied we must extend (a) from the sequence 2^n to all T. We still have some latitude in the above example, namely A and B are still somewhat arbitrary. Pick $A > 2$ and $1/4 < B < 1/2$ so that $A(1-B) < 3/2$ and choose $g_1 \geq 0$. We show by induction that

$\int_0^t f(s)ds \geq 2^n a_n$ on $[2^n, 2^{n+1}]$ if n is odd and $n \geq 3$ and

$\int_0^t f(s)ds \geq 2^{n-2} a_n$ on $[2^n, 2^{n+1}]$ if n is even. Note that since

$g_1 \geq 0$, $\int_0^t f(s)ds \geq 0$ on $[0,2]$. Now $g_2 \equiv 0$ so

$\int_0^t f(s)ds = \int_0^t g_1(s)ds$ on $[0,4]$ so we need to show, to get the induction started, that $\int_0^t g_1 \geq 2a_1 = 2$ on $[2,4]$, but

$\int_0^2 g_1 = 2m_1 = 2$ so this follows. Now suppose that $t \in (2^n, 2^{n+1})$.

One has $\int_0^t f(s)ds = \left(\int_0^{2^n} + \int_{2^n}^t\right) f(s)ds = 2^n a_n + \sum_{k=1}^n \int_{2^n}^t g_k(s)ds + \int_{2^n}^t g_{n+1}(s)ds$

$= 2^n a_n + \sum_{k=1}^n \int_0^{t-2^n} g_k(s)ds + \int_{2^n}^t g_{n+1}(s)ds$. The last equality is by the periodicity of g_k. Now since $g_m \equiv 0$ on $[0,2^n]$ if $m \geq n$, $\sum_{k=1}^n \int_0^{t-2^n} g_k(s)ds = \int_0^{t-2^n} f(s)ds$. We thus get the equation for $t \in (2^n, 2^{n+1})$

$$\int_0^t f(s)ds = 2^n a_n + \int_0^{t-2^n} f(s)ds + \int_{2^n}^t g_{n+1}(s)ds. \qquad (11.7)$$

If n is odd, then $m_{n+1} = 0$ so $g_{n+1} \equiv 0$. Thus $\int_0^t f(s)ds = 2^n a_n + \int_0^{t-2^n} f(s)ds \geq 2^n a_n$ since $\int_0^t f(s)ds \geq 0$ on $[0,2^n]$ by the induction hypothesis. If n is even, then we look at two cases. Case I is when $0 \leq t-2^n \leq 2^{n-1}$ and Case II $2^{n-1} \leq t-2^n \leq 2^n$. Note that for n even, $a_n = a_{n-1}$ so that in Case I, $\int_0^{t-2^n} f(s)ds \geq 0$ by the induction hypothesis, and in Case II, $\int_0^{t-2^n} f(s)ds \geq 2^{n-1}a_{n-1} = 2^{n-1}a_n$ by the induction hypothesis. Also

$$\int_{2^n}^t g_{n+1}(s)ds \geq -2^{n-1}\|g_{n+1}\| \geq -2^{n-1}A\,|m_{n+1}| \qquad (11.8)$$

where $m_{n+1} = a_{n+1} - a_n = (B-1)a_n$. So in Case I, using (11.7) we get $\int_0^t f(s)ds \geq 2^n a_n - 2^{n-1}A(1-B)a_n = 2^{n-1}a_n\,[2-A(1-B)] \geq 2^{n-1}a_n(\frac{1}{2})$.

In Case II, we get

$$\int_0^t f(s)ds \geq 2^n a_n + 2^{n-1}a_n - 2^{n-1}A(1-B)a_n = 2^{n-1}a_n[3-A(1-B)] \geq 2^{n-1}a_n .$$

In either case $\int_0^t f(s)ds \geq 2^{n-2}a_n$. This completes the construction of the function f.

Now consider the equation

$$x' = f(t)x \qquad (11.9)$$

with x scalar and f defined above. The solutions are $x(t) = x(t_0)\exp\left(-\int_{t_0}^t f(s)ds\right)$. For each t_0 fixed, $\int_{t_0}^t f(s)ds \to +\infty$ as $t \to +\infty$ so the zero solution is asymptotically stable. It is not uniformly asymptotically stable. To see this, note that if n is even, then $x(2^{n+1}) = x(2^n)\exp\left(-\int_{2^n}^{2^{n+1}} f(s)ds\right) \geq x(2^n)e^{(1-2B)2^n a_n}$ by (11.7) and (11.8). If δ and T are given, find n so that $2^n \geq T$, the above shows that $|x(2^n)| < \delta$ does not imply that $|x(t)| < \delta$ if $t \geq 2^n + T$.

Thus indirectly, Floquet Theory does not carry over to the almost periodic case. One can show directly that (11.9) is not kinematically similar to a constant. For suppose, there is a continuous $p(t)$ so that $0 < \epsilon_1 \leq p(t) \leq \epsilon_2 < \infty$ and $p' = cp + pf$. Then

$$p(t) = p(t_0)\exp\left(\int_{t_0}^t f\right)e^{c(t-t_0)} \quad \text{and} \quad |p(t)| = |p(t_0)|\exp\left(\int_{t_0}^t f\right)e^{Re(c)(t-t_0)}.$$

If $Re(c) > 0$, then $|p(t)| \to \infty$ as $t \to +\infty$. On the other hand, if $\mathrm{Re}(c) \leq 0$, then for $t_0 = 2^n$ and n even,

$$0 < \epsilon_1 \leq |p(2^{n+1})| \leq \epsilon_2 \exp\int_{2^n}^{2^{n+1}} f(s)ds \leq \epsilon_2 e^{(2\sqrt{B})^n(2B-1)} \to 0.$$

Either way one gets a contradiction.

To finish our round of examples, we also note that module containment is missing from Theorem 11.15. To show that it is not always true that the solutions have module in the module of the equation, we return to Example 10.5. Since the two solutions we described there do not have module containment, we need only show that they are uniformly asymptotically stable. This is easy to see, for if a solution is near φ_i, then $\xi_1^2 + \xi_2^2$ is near 1 and the scalar equations that the ξ_i's satisfy have stable solutions ± 1 and the equation is autonomous. Thus $\xi_1^2 + \xi_2^2 \to 1$ exponentially. If $\xi_1^2 + \xi_2^2$ is near 1 then $x_3' = x_3(1-2(x_1^2 + x_2^2))$ where $1 - 2(x_1^2 + x_2^2) < 0$. Again $x_3 \to 0$ exponentially. This shows that the φ_i are uniformly asymptotically stable.

5. <u>Quasi-stability in the large</u>. One of the applications that we will make is to certain Lienard equations which have the property that the distance between two solutions is monotone decreasing. This implies the uniform stability of a bounded solution. This is an instance where something can be said about solutions in the large. We introduce a type of stability which in some sense is weaker than uniform stability. On the other hand it is not a property that is localized.

<u>Definition 11.10</u>. Let φ be a solution of (11.1). Then φ is uniformly quasi-asymptotically stable in the large on $[t_0,\infty)$ if for every $\epsilon > 0$ and $r > 0$, there are numbers $T(r,\epsilon)$, $M(r)$ so that if $t_1 \geq t_0$ and $|x_1 - \varphi(t_1)| < r$, then

$$|x(t;t_1;x_1) - \varphi(t)| < M(r) \quad \text{for} \quad t \geq t_1 \tag{11.10}$$

and

$$|x(t;t_1;x_1) - \varphi(t)| < \epsilon \quad \text{for} \quad t \geq t_1 + T(\epsilon,r) . \tag{11.11}$$

The term "in the large" refers to the possibility of r being any positive number. One can restrict this to solutions in a given compact set. The property, abbreviated u.q.a.s.ℓ., is a semi-separating property in a very strong sense, and is inherited. To show this we prove lemmas which are analogues of Lemmas 11.8 and 11.14.

<u>Lemma 11.11</u>. Let (11.1) satisfy the standard hypothesis. If φ is a bounded u.q.a.s.ℓ. solution of (11.1) on $[t_0,\infty)$, $T_\alpha f(x,t) = g(x,t)$ and $T_\alpha\varphi = \psi$ uniformly on compact sets with $T_\alpha(\text{id}) = +\infty$, then ψ is a u.q.a.s.ℓ. solution of $x' = g(x,t)$ on R.

<u>Proof</u>: Let $\epsilon > 0$ be given along with $r > 0$, we may assume $0 < \epsilon < 1$. Note that ψ is defined on R, so let t_1 be a real number. If $|x_0 - \psi(t_1)| \leq r$, then for large n, $|x_0 - \varphi(t_1 + \alpha_n)| \leq r+1 = R$. Consider $x_f(t;t_1+\alpha_n;x_0)$. According to (11.10), there is a number $M(R)$ and $T(R,\epsilon/3) > 0$ so that

$|x_f(t+\alpha_n;t_1+\alpha_n;x_0) - \varphi(t+\alpha_n)| \le M(R)$ if $t \ge t_1$ and $|x_f(t+\alpha_n;t_1+\alpha_n;x_0) - \varphi(t+\alpha_n)| \le \epsilon/3$ if $t \ge t_1 + T(R,\epsilon/3)$, and if n is large, i.e. $t_1+\alpha_n \ge t_0$. If $y_n(t) = x_f(t;t_1+\alpha_n;x_0)$, then $|y_n(t)| \le M(R) + \|\varphi\|$ on $[t_1-\alpha_n,\infty)$ so that for some subsequence y_n converges uniformly on all compact subsets of R. Call the limit function y_0. Clearly y_0 is a solution of $x' = g(x,t)$ and $y(t_1) = x_0$. Thus $y(t) = x_g(t;t_1,x_0)$. If $t \ge t_1+T(R,\epsilon/3)$ is fixed, then $|y_n(t+\alpha_n) - y_0(t)| < \epsilon/3$ and $|\psi(t) - \varphi(t+\alpha_n)| < \epsilon/3$ for n large. Thus by the triangle inequality, $|y_0(t) - \psi(t)| < \epsilon$. A similar triangle inequality argument shows that $|y_0(t) - \psi(t)| \le \frac{2\epsilon}{3} + M(R) \le 1 + M(R) = \overline{M}(r)$ if $t \ge t_1$. Thus ψ is u.q.a.s.ℓ. on R.

Corollary 11.12. Under the standard hypothesis, u.q.a.s.ℓ. is an hereditary property.

Corollary 11.13. Under the standard hypothesis, if (11.1) has a bounded u.q.a.s.ℓ. solution on $[t_0,\infty)$ then it has a bounded u.q.a.s.ℓ. solution on R.

Lemma 11.14. Let $f(x,t)$ be almost periodic in t uniformly for x in compact sets. If there is a bounded u.q.a.s.ℓ. solution of (11.1) on R, then it is the only bounded solution on R.

Proof: Suppose φ is a bounded u.q.a.s.ℓ. solution and ψ is another bounded solution on R. Let $r > 0$ be such that $|\varphi(t) - \psi(t)| \le r$ for all t. Let $|\varphi(t_1) - \psi(t_1)| = \epsilon > 0$. If

$t_0 = t_1 + T(r,\epsilon/3)$, then we must have $|\varphi(t_1) - \psi(t_1)| < \epsilon/2$. This contradiction proves the result.

Note that we have stated and proved the preceding lemmas and corollaries as if we were looking at all of R^n, that is, $r > 0$ was arbitrary. However, the same proofs suffice to show that these results are also correct if we restrict ourselves to solutions in a given bounded set. In this way we get a slightly stronger theorem.

Theorem 11.15. Let (11.1) satisfy the standard hypothesis and have a bounded u.q.a.s.ℓ. solution on $[t_0,\infty)$. Then there is an almost periodic solution of (11.1) which is u.q.a.s.ℓ. on R, and the module of the solution is contained in the module of f.

Proof: By Corollary 11.13 we get a solution on R which is u.q.a.s.ℓ. on R, call it φ. Now each equation in the hull has exactly one solution in $\overline{\text{range}(\varphi)}$ which is u.q.a.s.ℓ. on R by Lemma 11.14 and Corollary 11.12. The result now follows from Theorem 10.1.

6. Total stability. From the physical standpoint, the most desirable kind of stability that we will talk about, is total stability. Roughly, this means that under small changes of initial conditions and equations, a certain solution does not change much.

Definition 11.16. A solution φ of $x' = f(x,t)$ is totally stable if for every $\epsilon > 0$, there is a $\delta > 0$ so that if $|\varphi(t_0) - x_0| < \delta$ and $|R(x,t)| < \delta$ on $K \times [t_0,\infty]$ and is continuous there, then $|\varphi(t) - x_{f+R}(t;t_0;x_0)| < \epsilon$ for $t \geq t_0$.

For our purposes we can restrict ourselves to $R(x,t)$ which are of a special form so it is worthwhile to record it as a special condition. It is the restriction of R to functions of the form $g-h$ for g and $h \in H(f)$. This is a quite natural condition arising out of the abstraction of differential equations to dynamical systems.

Definition 11.17. A solution $\varphi = \varphi_f(t;0;\bar{x})$ of (11.1) is stable under disturbances from the hull if for every $\epsilon > 0$ there is a $\delta > 0$ so that if $\tau \geq 0$, $\|f(x,t+\tau) - g(x,t)\| < \delta$, $g \in H(f)$, $x \in K$, and $\|\varphi_f(\tau;0;\bar{x}) - y\| < \delta$, then $|\varphi_f(t+\tau;0;\bar{x}) - \varphi_g(t;0;y)| < \epsilon$ for $t \geq 0$.

Note that $\varphi_f(t+\tau;0;\bar{x})$ is a solution of $\bar{x}' = f(\bar{x},t+\tau)$ $\bar{x}(0) = \varphi_f(\tau;0;\bar{x})$ whose initial condition is close to y, the initial condition of the nearby equation $x' = g(x,t)$. Also we do not intend that solutions to initial value problems are unique. That is, $\varphi_g(t;0;y)$ may represent any solution of $x' = g(x,t)$, $x(0) = y$.

By an easy change of variables, it follows that if φ is a solution of $x' = f(x,t)$ which is stable under disturbances of the hull and $\tau > 0$, then $\varphi(t+\tau)$ is a solution of $x' = f(x,t+\tau)$ that is stable under disturbances of the hull with the same ϵ,δ relationship. We use this observation in the next theorem.

Theorem 11.18. Let (11.1) be almost periodic in t uniformly for $x \in K$. If there is a solution φ in K that is stable under disturbances from the hull, then it is a.a.p. and its almost periodic part is also a solution.

Proof: Let $T_\alpha(id) = +\infty$ and $\varphi_k(t) = \varphi(t+\alpha_k)$. We may assume that

$T_\alpha f$ exists uniformly on $K \times R$. For large k,m we have $\|f(x,t+\alpha_k) - f(x,t+\alpha_m)\| < \delta$ on K. Also we may assume that $\{\varphi_k(t)\}$ converges uniformly on compact subsets of R^+. If $k \geq m$ and m is large, then $|\varphi_k(0) - \varphi_m(0)| < \delta$. Since φ_m is a solution of $x' = f(x,t+\alpha_m)$ that is stable under disturbances from the hull, we have $|\varphi_m(t) - \varphi_k(t)| \leq \epsilon$ for $t \geq 0$. That is, $T_\alpha\varphi$ exists uniformly on R^+. Thus φ is a.a.p. and its periodic part is also a solution. If $\varphi = p+q$, then note that in the above argument $T_\alpha q = 0$ uniformly on R^+ so it is easy to see that p is also stable under disturbances from the hull.

Corollary 11.19. If φ is a totally stable solution of (11.1) then φ is a.a.p. and its almost periodic part is a solution.

Proof: We merely observe that total stability implies stability under disturbances from the hull.

7. Uniform asymptotic stability. We see that total stability is a very strong property and it is rather easy to conclude that it implies an almost periodic solution. Surprisingly, total stability follows from apparently weaker hypothesis.

Lemma 11.20. Let $|f(x,t) - f(y,t)| \leq L|x-y|$ for $t \in R$ and $x,y \in K$. If φ is a uniformly asymptotically stable solution of (11.1) on $[0,\infty)$, then it is totally stable.

Proof: Let $\epsilon > 0$ be given and find δ and T from the uniform asymptotic stability so that if $t_0 > 0$ and $|\varphi(t_0) - x_0| < \delta$, then $|\varphi(t) - x_f(t;t_0;x_0)| < \epsilon/2$ if $t \geq t_0$ and $|\varphi(t) - x_f(t;t_0;x_0)| < \delta/2$ if $t \geq t_0 + T$. Choose $\delta < \epsilon$. Let $\delta_1 = \delta/2\, L(e^{LT}-1)^{-1}$. If ψ is a solution of $x' = f(x,t) + R(x,t)$ with R continuous $|R(x,t)| < \delta_1$, and $\psi(t_0) = x_0$, we have
$\psi(t) = x_0 + \int_{t_0}^{t} f(\psi(s),s)ds + \int_{t_0}^{t} R(\psi(s),s)ds$. Furthermore,

$x_f(t;t_0;x_0) = x_0 + \int_{t_0}^{t} f(x_f(s;t_0;x_0),s)ds$ so for $t \geq t_0$, we have

$|x_f(t;t_0;x_0) - \psi(t)| \leq \int_{t_0}^{t} L\, |x_f(s;t_0;x_0) - \psi(s)|ds + (t-t_0)\delta_1$. By a version of Gronwall's inequality,

$|x_f(t;t_0;x_0) - \psi(t)| \leq \frac{\delta_1}{L}[e^{L(t-t_0)}-1] < \delta/2$ if $0 \leq t-t_0 \leq T$.

Thus $|\psi(t) - \varphi(t)| \leq \delta/2 + \epsilon/2 < \epsilon$ for $t_0 \leq t \leq t_0 + T$.

Now note that $|\varphi(t_0+T) - \psi(t_0+T)| < \delta/2 + \delta/2 = \delta$ so that $\psi(t_0+T) = x_1$, satisfies $|x_1 - \varphi(t_0+T)| < \delta$ so one can repeat the above argument with x_0 replaced by x_1 to get $|\psi(t+T) - \varphi(t+T)| < \epsilon$ on $(t_0,t_0+T]$. That is, $|\varphi(t) - \psi(t)| < \epsilon$ on $[t_0,t_0+2T]$. Now we do an obvious induction to conclude that $|\varphi(t) - \psi(t)| < \epsilon$ on $[t_0,\infty)$, so it is totally stable.

Corollary 11.21. Let $f(x,t)$ be almost periodic in t for each $x \in K$ and satisfy $|f(x,t) - f(y,t)| \leq L|x-y|$ for $x,y \in K$ and $t \in R$. If φ is a uniformly asymptotically stable solution to (11.1) in K, then φ is a.a.p. and its almost periodic part is a uniformly asymptotically stable solution of (11.1).

Proof: This is mostly calling forth previous results. Firstly, φ is totally stable. Secondly, f satisfies the standard hypothesis so φ is a.a.p. If $\varphi = p+q$, then we know p is a solution. But we know more. $T_\alpha \varphi = p$ if $T_\alpha(\text{id}) = +\infty$. If further $T_\alpha f = f$, then p is a uniformly asymptotically stable solution, since that property is inherited.

The crucial part of the above proof is the use of the Lipschitz condition to get a uniform estimate for $|x_f(t;t_0;x_0) - \psi(t)| < \delta/2$ on $[t_0,t_0+T]$ under the assumption that $\psi(t_0) = x_0$, and ψ is $x_{f+R}(t;t_0;x_0)$ where $|R| < \delta_1$. If one works slightly harder, we can prove this under the slightly weaker hypothesis that (11.1) satisfies the standard hypothesis. Then the Corollary 11.21 has the same extension. This proves to be a satisfactory answer to the question of almost periodicity following from uniform asymptotic stability.

The next lemma is a slight generalization of Lemma 9.6.

Lemma 11.22. Let (11.1) satisfy the standard hypothesis, and φ be a solution in K on $(0,\infty)$. If T and $\epsilon > 0$ are given, there is a $\delta > 0$ so that if $|R(x,t)| < \delta$ and $t_0 \geq 0$, then $|\varphi(t) - x_{f+R}(t;t_0;\varphi(t_0))| < \epsilon$ on $[t_0,t_0+T]$.

Proof: We assume not. There are an $\epsilon > 0$, $\delta_k \downarrow 0$, $t_k > 0$, $|R_k(x,t)| < \delta_k$ and $\tau_k \in [t_k,t_k+T]$ such that $|\varphi(t_k) - x_{f+R_k}(t;t_k;\varphi(t_k))| < \delta_k$, $|\varphi(\tau_k) - x_{f+R_k}(\tau_k;t_k;\varphi(t_k))| = \epsilon$ and

$|\varphi(t) - x_{f+R_k}(t;t_k;\varphi(t_k))| < \epsilon$ on $[t_k, \tau_k)$. Necessarily, $t_k \to +\infty$ by Kamke's Theorem. For convenience, write $x_k(t) = x_{f+R_k}(t;t_k;\varphi(t_k))$. We take subsequences so that $\sigma_k = \tau_k - t_k$ converges to σ_0. This is possible since $0 \leq \sigma_k \leq T$. Also define $\varphi_k(t) = \varphi(t+t_k)$ and $\psi_k(t) = x_k(t+t_k)$. Then φ_k is a solution of $x' = f(x,t+t_k)$ and ψ_k a solution of $x' = f(x,t+t_k) + R_k(x,t+t_k)$. Further, $\varphi_k(0) = \psi_k(0)$. By taking further sequences, we get $T_t f = g$ uniformly, and $T_t\varphi_k = \varphi^*$ and $T_t\psi_k = \psi^*$ uniformly on compact sets. Then φ^* and ψ^* are solutions of $x' = g(x,t)$ on $[0,T]$ with $\varphi^*(0) = \psi^*(0)$. Thus $\varphi^* \equiv \psi^*$. But $|\varphi^*(\sigma_0) - \psi^*(\sigma_0)| = \lim_k |\varphi_k(\sigma_0) - \psi_k(\sigma_0)|$

$= \lim_k |\varphi(t_k + \sigma_k) - x_k(t_k + \sigma_k)| = \lim_k |\varphi(\tau_k) - x_k(\tau_k)| = \lim_k \epsilon = \epsilon > 0.$

Corollary 11.23. Let (11.1) satisfy the standard hypothesis. If φ is a bounded uniformly asymptotically stable solution on $[t_0,\infty)$, then φ is a.a.p. and its almost periodic part is a uniformly asymptotically stable solution.

Proof: As we noted above, φ is totally stable by an argument analogous to that in Lemma 11.20, where the Lipschitz condition is replaced by Lemma 11.22. Then the proof is finished as in Corollary 11.21.

An interesting question arises.

Question K: Are Corollaries 11.21 and 11.23 true if uniform asymptotic stability is replaced by uniform stability? Perhaps, if φ is not a.a.p. some other solution is.

8. Periodic Equations. The importance of question K above, is that when the results of the previous section are applied to periodic equations, do we get periodic solutions? If the answer to K is yes, then we do. Thus Theorem 11.15 does give periodic solutions if f is periodic. It does not, however guarantee that the period of the solution is the same as the period of the equation.

We would like to talk about periodic systems here since some of the results of the previous section can be improved. Specifically, we may sometimes drop the hypothesis of unique solutions to initial value problems. The reason this is so, is that a solution of an equation in the hull is merely the translate of a solution of the original equation so no limit argument is required for the hereditary property.

Theorem 11.24. Let (11.1) be periodic and have a bounded uniformly stable solution φ. Then φ is a.a.p. and its almost periodic part is a solution.

Proof: Take α' so that $T_{\alpha'}(id) = +\infty$. Write $\alpha'_k = \eta'_k\omega + \sigma'_k$ for $0 \leq \sigma'_k < \omega$, where ω is the period of f. Take $\alpha \subset \alpha'$ so that $T_\alpha\varphi$ exists uniformly on compact subsets of R^+ and so that the common subsequence $\sigma \subset \sigma'$ converges to σ_0. Then $T_\alpha f = T_\alpha f = T_{\sigma_0} f$. Define $\varphi_k(t) = \varphi(t+\alpha_k)$ which is a uniformly stable solution of

$x' = f(x,t+\sigma_k)$ (same δ and ϵ as φ). Furthermore, let $|x-y| < \eta$ imply that $|\varphi_k(x) - \varphi_k(y)| < \epsilon$ for all k. Consider now the two functions $\varphi_k(t)$ and $\varphi_m(t+\alpha_k - \sigma_m)$ with $m \geq k$. They are both solutions of $x' = f(x,t+\sigma_k)$ and if k is sufficiently large, $|\varphi_k(0) - \varphi_m(\sigma_k - \sigma_m)| \leq |\varphi_k(0) - \varphi_m(0)| + |\varphi_m(0) - \varphi_m(\sigma_k - \sigma_m)| < \delta$. By the uniform stability of φ_k we have $|\varphi_k(t) - \varphi_m(t+\sigma_k - \sigma_m)| < \epsilon$ for $t \geq 0$ if k is sufficiently large. On the other hand, if k is large then $|\sigma_k - \sigma_m| < \eta$ so that $|\varphi_m(t) - \varphi_m(t+\sigma_k - \sigma_m)| < \epsilon$ for all $t \geq 0$. Combining these two, we get $|\varphi_k(t) - \varphi_m(t)| < 2\epsilon$ for $t \geq 0$, and $m \geq k$ sufficiently large. That is $T_\alpha \varphi$ exists uniformly on R^+, and φ is a.a.p.

<u>Question L</u>. Must the almost periodic solution be periodic?

The other natural question that arises, is the uniform stability of the almost periodic solution. This is answered in the affirmative by the next lemma. Note the absence of uniqueness of solutions of initial value problems.

<u>Lemma 11.25</u>. For periodic systems, uniform stability and uniform asymptotic stability are inherited.

<u>Proof</u>: Let φ be uniformly stable and $T_\alpha(id) = +\infty$ with $T_\alpha \varphi = \psi$ uniformly on compact subsets of R and $\alpha_k = n_k \omega + \sigma_k$ with σ_k converging to σ_0. Let $\epsilon > 0$ and choose δ as in uniform stability, then pick $t_0 \in R$, and x_0 such that $|\psi(t_0) - x_0| < \delta$. Now ψ is a solution of $x' = f(x,t+\sigma_0)$. Define η by $|\psi(t_0) - x_0| < \eta < \delta$,

and $\varphi_k(t) = \varphi(t+\alpha_k)$. For large k, $|\varphi_k(t_0) - \psi(t_0)| < \frac{\delta-\eta}{2}$, and $|\varphi(t_0+\sigma_0+n_k\omega) - \varphi(t_0+\sigma_k+n_k\omega)| < \frac{\delta-\eta}{2}$, the latter by uniform continuity. Then by the triangle inequality, $|\varphi(t_0+\sigma_0+n_k\omega) - x_0| < \delta$. Now $\varphi(t+\sigma_0+n_k\omega)$ is a uniformly stable solution of $x' = f(x,t+\sigma_0)$ with same ϵ,δ as φ. Let $x(t)$ be a solution of this equation through (t_0,x_0). Thus $|\varphi(t+\sigma_0+n_k\omega) - x(t)| < \epsilon$ for $t \geq t_0$. But for k large we also have $|\varphi(t_0+\sigma_0+n_k\omega) - \psi(t_0)| < \delta_1$ where δ_1 is the δ from uniform stability belonging to η. Thus $|\varphi(t+\sigma_0+n_k\omega) - \psi(t)| < \eta$ for $t \geq t_0$. Combining we get $|\psi(t) - x(t)| < \epsilon + \eta$ for $t \geq t_0$. But $\eta < \delta$ so $|\psi(t) - x(t)| < \epsilon + \delta$ for $t \geq t_0$. Uniform asymptotic stability follows in the same way.

Corollary 11.26. The almost periodic solution of Theorem 11.24 is uniformly stable.

We can now strengthen the results of Theorem 11.24 if the solution is uniformly asymptotically stable.

Theorem 11.27. If (11.1) is periodic and has a bounded uniformly asymptotically stable solution on $[t_0,\infty)$, then there is a periodic solution which is uniformly asymptotically stable.

Proof: Let φ be the solution that is uniformly asymptotically stable, and let $\alpha'_k = k\omega$. Then φ is a.a.p. by Theorem 11.24 so there is $\alpha \subset \alpha'$ so that $T_\alpha\varphi$ exists uniformly on R^+. Let $r > 0$ be given by Definition 11.7 and observe that $\varphi(t+k\omega)$ has the same r, and T as φ. There are k_1 and k_2 so that

$|\varphi(\alpha_{k_1}) - \varphi(\alpha_{k_2})| < r$. Consequently, $|\varphi(t+\alpha_{k_1}) - \varphi(t+\alpha_{k_2})| \to 0$ as $t \to +\infty$, since both are solutions of (11.1) and are uniformly asymptotically stable. It follows that $|\varphi(t + \alpha_{k_1} - \alpha_{k_2}) - \varphi(t)| \to 0$ as $t \to +\infty$. But $\varphi(t) = p(t) + q(t)$ where p is almost periodic and $\lim_{t\to\infty} q(t) = 0$. It follows that $\lim_{t\to\infty} |p(t + \alpha_{k_1} - \alpha_{k_2}) - p(t)| = 0$. For almost periodic functions this implies that $p(t + \alpha_{k_1} - \alpha_{k_2}) \equiv p(t)$. Note that $m = \alpha_{k_1} - \alpha_{k_2}$ is an integral multiple of ω.

One cannot conclude that $m = 1$ in view of Example 10.5.

<u>Corollary 11.28</u>. If (11.1) is periodic and φ is a bounded solution that is u.q.a.s.ℓ., then φ is periodic with the same period as f.

<u>Proof</u>: We already know that φ is a.a.p. since it is uniformly stable. But $\varphi(t+\omega)$ is another bounded solution, so $\varphi(t+\omega) = \varphi(t)$ by Lemma 11.14.

In this connection, it is possible to have a uniformly stable solution to an almost periodic system so that all solutions are asymptotic, but that solution is not u.q.a.s.ℓ.

<u>Example 11.29</u>. Consider the scalar equation

$$x' = g(x,t) = \begin{cases} -x & \text{if } 0 \le x \le 1 \\ -1 + (1-2f(t))(x-1) & \text{if } 1 \le x \le 2, \\ -x\, f(t) & \text{if } x > 2, \\ \text{odd in } x. & \end{cases}$$

where $f(t)$ is the function described in Example 11.19. We do not do the details, they are standard geometric arguments. The reason that u.q.a.s.ℓ. fails is that, if $x_n > 2$ then $x(t;t_n;x_n)$ satisfies $x(t_n + S_n; t_n;x_n) > x_n e^{nS_n}$ where t_n and S_n are defined in 11.19.

9. Perturbations of stable systems. We do not intend to give a description of a general theory of perturbations of stable systems. There are the results using Lyapunov functions which may be found in the monograph on that subject by Taro Yoshizawa. We will not reproduce those here, nor the theorems that come from global analysis. We are content to discuss a theory of perturbations of non-linear equations that are exponentially stable. For linear systems this means we will need an exponential dichotomy with $P = I$.

Specifically assume that

H: (i) $x' = A(t)x$ (11.12)

satisfies an exponential dichotomy $|X(t)X^{-1}(s)| \leq Ke^{-\alpha(t-s)}$, with $A(t)$ almost periodic

(ii) $f(0,t) = 0$ and $|f(x,t) - f(y,t)| \leq L|x-y|$ for all x and t, with $f(x,t)$ almost periodic in t.

(iii) $g(x,t)$ is differentiable and satisfies the standard hypothesis with $g(x,t) = o(|x|)$ as $x \to 0$ uniformly in t.

(iv) $h(t) \in AP(E^n)$.

Along with the above we define exponential stability.

Definition 11.30. A solution φ of a differential equation is exponentially stable if φ is bounded on R and there is an $r > 0$ so that any other solution ψ which satisfies $|\psi(t_0) - \varphi(t_0)| < r$, also satisfies $|\psi(t) - \varphi(t)| \leq |\psi(t_0) - \varphi(t_0)| M(r)^{\sigma(r)}$ for $t \geq t_0$.

If $r = +\infty$, it is clear that exponential stability implies uniform asymptotic stability and u.q.a.s.ℓ. Also the system (11.12) has the given exponential dichotomy if and only if the zero solution is exponentially stable with $r = +\infty$. We also remark that by H(iii) g satisfies a Lipschitz condition near $x = 0$ that is uniform in t. This Lipschitz constant shrinks to zero as the region near zero shrinks to zero. We now consider the equation

$$x' = A(t) + f(x,t) + g(x,t) + h(t). \tag{11.13}$$

<u>Theorem 11.31</u>. Let (11.13) satisfy H(i), (ii), (iii) and (iv). If $L < K^{-1}\alpha$, then there is a B_0 so that if $\|h\| < B_0$, then (11.13) has an almost periodic exponentially stable solution whose module is contained in the module of (11.13).

<u>Proof</u>: We use the contraction mapping principle. Let ϵ be chosen so that $L + \epsilon < K^{-1}\alpha$ and δ chosen so that the Lipschitz constant of g is less than ϵ on the ball $\|x\| \leq \delta$. Define $B_0 = (K^{-1}\alpha - L - \epsilon)\delta$, and consider the space $\mathfrak{B}$ of functions in $AP(E^n)$ with norm less than or equal to δ. For $\varphi \in \mathfrak{B}$, define

$$(T\varphi)(t) = -\int_t^\infty X(t)X^{-1}(s)[f(\varphi(s),s) + g(\varphi(s),s) + h(s)]ds. \tag{11.14}$$

Note that since $f(0,t) = g(0,t) = 0$ for all t, we have $|f(\varphi(s),s)| \le L|\varphi(s)|$ and $|g(\varphi(s),s)| \le \epsilon|\varphi(s)|$. It follows that

$$\|T\varphi\| \le \int_t^\infty K e^{\alpha-(t-s)} [(L+\epsilon)|\varphi(s)| + \|h\|]ds < K(L\delta + \epsilon\delta + B_0)\alpha^{-1} = \delta.$$

Thus T maps $\mathfrak{B}$ into itself since $T\varphi$ is a bounded solution to $x' = A(t)x + f(\varphi(t),t) + g(\varphi(t),t) + h(t)$ and is thus almost periodic by Theorem 7.7. Similarly $\|T\varphi - T\psi\| \le \int_t^\infty K e^{-\alpha(t-s)} (L+\epsilon)\|\varphi-\psi\|ds$ $= K\alpha^{-1}(L+\epsilon)\|\varphi-\psi\|$ so is a contraction. Let φ now be the fixed point of T. We want to show that φ is exponentially stable. Let ψ be a solution such that $|\varphi(t_0) - \psi(t_0)| < K^{-1}(\delta - \|\varphi\|)$. This implies that $|\psi(t_0)| < \delta$ so by continuity there is an interval $[t_0,t_1)$ where $|\psi(t)| < \delta$. On this interval, ψ has the representation

$$\psi(t) = \int_{t_0}^t X(t)X^{-1}(s)[f(\psi(s),s) + g(\psi(s),s) + h(s)]ds + X(t)X^{-1}(t_0)\psi(t_0)$$

as does φ. Thus

$$|\psi(t) - \varphi(t)| \le K e^{-\alpha(t-t_0)} |\psi(t_0) - \varphi(t_0)| + \int_{t_0}^t K e^{-\alpha(t-s)} [(L+\epsilon)|\psi(s)-\varphi(s)|ds.$$

By Gronwall's inequality,

$$|\psi(t) - \varphi(t)| \le |\psi(t_0) - \varphi(t_0)| K e^{-(\alpha-K(L+\epsilon))(t-t_0)} \tag{11.14}$$

if $t_0 \le t \le t_1$. But then $|\psi(t) - \varphi(t)| \le (\delta - \|\varphi\|) < \delta$ on this interval. Now it is clear that we may take $t_1 = +\infty$ and (11.14) shows the exponential stability. To get module containment one merely observes that φ is a fixed point of the mapping T where the functions in $\mathfrak{B}$ are restricted to have module in the module of (11.13). This follows since T is a solution to an equation of the form

$x' = A(t)x + f(t)$ where f has the module containment and so Tf does too, again see Theorem 7.7.

Corollary 11.31. Let H(i), (ii) and (iv) be satisfied with $L < K^{-1}\alpha$. Then the equation

$$x' = A(t)x + f(x,t) + h(t) \tag{11.15}$$

has an almost periodic exponentially stable solution with module in module of (11.15).

Proof: We note that the proof of Theorem (11.30) works here, but we may take δ arbitrary since $g \equiv 0$. Thus B_0 may be taken arbitrarily large and $\|h\|$ is unrestricted.

There is a slight generalization of the above results. If the zero solution of $x' = A(t)x + g(x,t)$ is exponentially stable one also gets the above results. The idea is that one can deduce from this that H(i) is satisfied. We will not do the details.

10. Strong stability again. We began this chapter with the theorem that bounded strongly stable motions are almost periodic. We have exposed how a great many weaker stability conditions also yield almost periodic solutions. What is surprising now is that strong stability is necessarily true for almost periodic equations. Hence we have come the full circle.

Theorem 11.32. Let (11.1) satisfy the standard hypothesis. Then any almost periodic solution is strongly stable.

Proof: Suppose not. Then there are sequences α,β,γ and an $\epsilon > 0$ so that $|\varphi(\alpha_n) - \varphi(\beta_n)| < \frac{1}{n}$ and $|\varphi(\alpha_n + \gamma_n) - \varphi(\beta_n + \gamma_n)| \geq \epsilon$. By taking common subsequences we may assume that $T_\gamma T_\alpha \varphi = T_{\alpha+\gamma}\varphi$ and $T_\gamma T_\beta \varphi = T_{\beta+\gamma}\varphi$ and the same equations for f. Now $T_\alpha\varphi(0) = T_\beta\varphi(0)$ and by uniqueness of initial value problems, $T_\alpha\varphi \equiv T_\beta\varphi$. Thus $T_{\alpha+\gamma}\varphi \equiv T_{\beta+\gamma}\varphi$ and on the other hand, $|T_{\alpha+\gamma}\varphi(0) - T_{\beta+\gamma}\varphi(0)| \geq \epsilon$.

11. Notes. The notion of Δ-mset and Theorem 11.5 are in Markov, Stabilität in Liapvanoffschen Sinne und Fastperiodizität, Math. Z. 36(1933), 708-738. More on Δ-m sets may be found in Meisters and Fink, On Δ-m sets and group topologies, Rocky Mount. J. 2(1972), 225-233.

The notions of stability are standard. They can be found on any text that discusses Lypanov's second method. See for example, Yoshizawa, Stability theory by Liaponov's second method, Math. Soc. of Japan, 1966. Among other things, sufficient conditions for stability are given, as well as many examples.

Theorem 11.15 is in Fink, Semi-separated conditions for almost periodic solutions, J. Differential Equations, 11(1972), 245-251. An earlier version allowing only one uniformly stable solution is given in Fink and Seifert, Liaponov Functions and Almost periodic solutions for almost periodic systems, J. Differential Equations, 5(1969), 307-313.

The construction of examples is very important. The examples 11.17 and 11.18, given here are found in (respectively)

Opial, Sur les solutions presque-periodiques d'une classe d'equations differentielles, Ann. Polon. Math. 9(1960/61), 157-181; and Sell, Nonautonomous differential equations and topological dynamics I., Trans. Amer. Math. Soc. 127(1967), 241-262, and Kato, Uniform asymptotic stability and total stability, Tohoku Math. J. 22(1970), 254-269; with example 11.19 in Conley and Miller, Asymptotic stability without uniform stability: almost periodic coefficients, J. Differential Equations, 1(1965), 333-336.

The theorems on u.q.a.s.ℓ. are in Seifert, Almost periodic solutions and asymptotic stability, J. Math. Anal. Appl. 21(1968), 136-149. Theorem 11.18, Corollary 11.19 and Lemma 11.20 are in Coppel, Almost periodic properties of ordinary differential equations, Ann. Mat. pura appl. 76(1967), 27-50, and Miller, Almost periodic differential equations as dynamical systems with applications to the existence of almost periodic solutions, J. Differential Equations 1(1965), 337-345.

Theorem 11.24 is due to Deysach and Sell, On the existence of almost periodic motions, Michigan Math. J. 12(1965), 87-95, while theorem 11.27 is in Sell, Periodic solutions and asymptotic stability, J. Differential Equations 2(1966), 143-157.

Example 11.29 is in Seifert, Almost periodic solutions and asymptotic stability, J. Math. Anal. Appl. 2(1968), 136-149.

The results of article 9 are in Sell, Some perturbation problems in ordinary differential equations, Funkcialaj Ekvacioj 10(1967), 1-13. Theorem 11.32 is due to Franklin, Almost periodic recurrent motions,

Math. Z. 30(1929), 325-331.

For further results on stability and almost periodicity, the papers of Seifert, Yoshizawa and Sell listed in Chapter 15 are recommended.

The organization of the results of this chapter and the gist of their proofs is an amalgamation of the two papers, Yoshizawa, Asymptotically almost periodic solutions of an almost periodic system, Funkcialaj Ekvacioj 12(1969), 23-40, and Fink, Semi-separated conditions for almost periodic solutions, J. Differential Equations, 11(1972), 245-251.

The following example answers Question J in the affirmative.

Example (due to George Seifert). Let $a(t)$ be real, almost periodic, with mean value zero, but unbounded integral. Consider the real system

$$x_1' = -a(t)x_2$$

$$x_2' = a(t)x_1$$

$$x_3' = 1 - (x_1^2 + x_2^2)x_3$$

which can be written as

$$x' = i\,a(t)x$$

$$x_3' = 1 - |x|^2 x_3$$

with $x = x_1 + i\,x_2$.

The solutions are

$$x(t) = x_0(t_0)\exp i\left(\int_{t_0}^{t} a(s)ds\right)$$

$$x_3(t) = x_3(t_0)e^{-|x(t_0)|^2(t-t_0)} + [1 - e^{-|x(t_0)|^2(t-t_0)}]\,|x(t_0)|^{-2}$$

$$\text{if } x(t_0) \neq 0$$

and

$$x_3(t) = x_3(t_0) + (t - t_0) \qquad \text{if} \quad x(t_0) = 0 .$$

If $x_3(t_0) = |x(t_0)|^{-2}$, then $x_3(t) \equiv |x(t_0)|^{-2}$. Therefore any such solution is bounded on $\mathbb{R}$ and clearly uniformly stable. Since $x(t)$ is of constant norm, no solution is asymptotically stable.

Chapter 12

First order equations

1. Introduction. Many of the techniques and conjectures that arise in almost periodic differential equation theory are motivated by analogous results from periodic differential equations. Many interesting ones appear already in the scalar case. We are going to give a few positive results, but mostly we give examples to show that things are quite difficult already in the scalar case.

2. Periodic equations. The easiest sufficient condition for the existence of a periodic solution is merely the existence of a unique bounded solution. For if φ is a solution of $x' = f(x,t)$ and $f(x,t+\omega) = f(x,t)$, then $\varphi(t+\omega)$ is another bounded solution so $\varphi(t+\omega) = \varphi(t)$. Of course the same hypothesis for almost periodic equations is an interesting one and it is unknown whether such a solution need be almost periodic, see Question I.

For periodic scalar equations, the existence of a solution bounded on $[0,\infty)$ implies the existence of a periodic solution.

Theorem 12.1. Suppose $x' = f(x,t)$ is a scalar equation with unique solutions of initial value problems. If $f(x,t+\omega) = f(x,t)$ and φ is a solution bounded on $[0,\infty)$, then there is a solution of period ω.

Proof: We consider the solutions $x_n(t) = \varphi(t+n\omega)$. They are bounded on $[0,\omega]$. If φ is not periodic of period ω, then

$x_1(0) = \varphi(\omega) \neq \varphi(0)$, say $\varphi(\omega) > \varphi(0)$. Then $x_1(t) > x_o(t)$ on $[0,\omega]$ since solutions may not cross. In particular $x_2(0) = x_1(\omega) > x_o(\omega) = x_1(0)$ and thus $x_2(t) > x_1(t)$. By induction, $x_{n+1}(t) > x_n(t)$ on $[0,\omega]$. Since φ is bounded, $x_n(0)$ is a bounded monotone increasing sequence so converges to x_o. Thus $x_n(t) \to x_o(t)$ where x_o is a solution of $x' = f(x,t)$ and $x_o(0) = x_o$. Now $x_o(\omega) = \lim_{n\to\infty} x_n(\omega) = \lim_{n\to\infty} x_{n+1}(0) = x_o(0)$ so x_o is periodic.

The result corresponding to Theorem 12.1 if $x \in E^n$, $n > 1$ is not true, but all known examples where all solutions are bounded have almost periodic solutions. In the scalar case, it is not possible to have almost periodic solutions and no periodic solutions.

<u>Theorem 12.2</u>. If the scalar periodic equation $x' = f(x,t)$ has an almost periodic solution, then it is a periodic solution. (Assuming unique solutions of initial value problems.)

<u>Proof</u>: Let φ be an almost periodic solution. Consider $\alpha_n' = n\omega$ and find $\alpha \subset \alpha'$ so that $T_\alpha \varphi = \psi$ uniformly, then $\varphi(t+\alpha_n)$ is a solution of $x' = f(x,t)$ and as in Theorem 12.1, $\varphi(n\omega)$ is monotone so in fact $T_{\alpha'}\varphi = \psi$. Thus as in Theorem 12.1, $\psi(0) = \psi(\omega)$. Now $\varphi = T_{-\alpha'}\psi$ is also periodic.

3. <u>Differential Equations on a Torus</u>. We will devote some time in discussing some interesting equations which arise from assuming that $f(x,t)$ is periodic in both variables. One may

view the equation $x' = f(x,t)$ as being defined on a torus. We will not develope the theory in any comprehensive way but refer any discussion of details to the standard treatises. In particular, we refer to Chapter 17 of the book by Coddington and Levinson and lift out the facts that we shall need from that discussion.

Specifically consider the equation

$$x' = f(x,t) \tag{12.1}$$

with $f(x+1,t) = f(x,t+1) = f(x,t)$, and f continuous and such that (12.1) has unique solution to initial value problems.

Let $\varphi(t,\eta)$ be the solution such that $\varphi(0,\eta) = \eta$ and $\psi(\eta) = \varphi(1,\eta)$. Then ψ is a homeomorphism of R and we may iterate. The average rate of gain of ψ namely

$$\rho = \lim_{n\to\infty} \frac{\psi^n(\eta)}{n} \tag{12.2}$$

exists for each η and is independent of η. It is called the rotation number because of its interpretation on the torus. A particular estimate for ρ that we will use is

$\left|\frac{\psi^n(\eta)}{n} - \frac{\psi^n(0)}{n}\right| \leq \frac{1}{n}$. This estimate occurs in the proof of (12.2).

Another is the estimate $\left|\frac{\psi^n(0)}{n} - \rho\right| \leq \frac{2}{n}$. Combining these two we have $\left|\frac{\psi^n(\eta)}{n} - \rho\right| \leq \frac{3}{n}$ so that $|\psi^n(\eta) - \rho n| \leq 3$ (12.3)

This equation shows that $|\varphi(n,\eta) - \rho n|$ is at most 3 so that $\varphi(t,\eta) - \rho t$ is uniformly bounded on R.

Two cases arise which are distinguished by the character of

ρ. If ρ is rational, then as a flow on the torus, there is a periodic solution and corresponds to the existence of an integer m so that $\omega(m,\eta) - m\rho = 0$. Conversely, if the flow on the torus has a periodic solution, then ρ is rational.

The more interesting case is when ρ is irrational. Let $S \equiv \{\psi^n(\eta) \mod 1;\ n \in \text{integers}\}$, and S' be its limits points. Then S' does not depend on η and is invariant under ψ, that is, if $\eta \in S'$ then $\psi(\eta) \mod 1 \in S'$. Furthermore, S' is a closed set and dense in itself, that is every point in S' is a limit point of the rest of S'. Now only two possibilities arise for S'. Either $S' = [0,1]$ or S' is nowhere dense. The latter is a "Cantor set". The first case is called the ergodic case and the second is the singular case. The next theorem is one of the few theorems in which differentiability plays an essential role. In most cases, continuity and differentiability are more related to methods of proof then essential to the truth of the theorem.

Theorem 12.3. If f is continuous and $\frac{\partial f}{\partial x}$ is continuous and a function of bounded variation in x uniformly in t, then the singular case does not arise. That is, there is a periodic solution on the torus, or $S' = [0,1]$. Furthermore, there is an f for which $\frac{\partial f}{\partial x}$ is continuous and the singular case occurs.

In the ergodic case, there is a nice representation of the solutions of (12.1).

Theorem 12.4. If (12.1) is in the ergodic case, then there is a continuous function $w(x,t)$ such that

$w(x+1,t) = w(x,t+1) = w(x,t)$ and if φ is a solution of (12.1) then

$$\varphi(t) = \rho t + c + w(t,\rho t + c) \tag{12.4}$$

where ρ is the (irrational) rotation number and c is an appropriate constant. Conversely, for every c, the right hand side of (12.4) is a solution of (12.1).

It is easy to see that the function $w(t,\rho t + c)$ is almost periodic so that $\varphi(t) - \rho t$ is not only uniformly bounded, but also almost periodic. Furthermore, it is a solution to the equation $x' = f(x+\rho t,t) - \rho$ which is an almost periodic equation. This is an example of a non-linear almost periodic equation all of whose solutions are bounded and almost periodic. There is an example at the other end of the spectrum.

Example 12.5. There is a scalar almost periodic differential equation all of whose solutions are bounded but none are almost periodic. We begin with a differential equation (12.1) in the singular case. This means the set S' is nowhere dense. We also assume unique solutions to initial value problems. Since the set S' is dense in itself, for every η_o and $\eta_1 \in S'$, there are sequences of integers q_n, p_n such that $\varphi(p_n,\eta_o) + q_n \to \eta_1$. By continuous dependence of solutions of initial value problems, $\varphi(t+p_n,\eta_o) + q_n \to \varphi(t,\eta_1)$ on compact sets. Now as before $y(t) = \varphi(t,\eta) - \rho t$ is a bounded solution of the almost periodic differential equation

$$x' = f(x+\rho t,t) - \rho \tag{12.5}$$

and this correspondence between solutions is 1 to 1 and onto. To show that no solution is almost periodic we first show that if some solution is almost periodic, then all solutions beginning in S' are almost periodic. For if $x(t;0;\eta_o)$ is an almost periodic solution of (12.5), then let $\eta_1 \in S'$ be arbitrary. By the definition of S' and the independence of S on η, we can find integers p'_n and q'_n so that $\varphi(t+p'_n,\eta_o) + q'_n \to \varphi(t,\eta_1)$. Since $x(t;0;\eta_o)$ is almost periodic, we can find $p \subset p'$ and $q \subset q'$ common subsequences, so that $x(t+p_n;0;\eta_o) \to y(t)$ an almost periodic function. Now $x(t+p_n;0;\eta_o) = \varphi(t+p_n,\eta_o) - \rho(t+p_n)$ $= \varphi(t+p_n,\eta_o) + q_n - \rho t - (q_n + \rho p_n)$. We may take limits to get $y(t) = \varphi(t,\eta_1) - \rho t + \beta$, where $\beta = \lim\limits_{n\to\infty} - (q_n + \rho p_n)$ exists since all other limits in the equation exist. Thus $y(t) = x(t;0;\eta_1) + \beta$ and $x(t;0;\eta_1)$ is almost periodic.

We now show that it is not possible to have all solutions $\{x(t;0;\eta) : \eta \in S'\}$ almost periodic. Since S' is nowhere dense and closed, the complement of S' is a countable union of disjoint open intervals (a_n,b_n). Let (η_1,η_2) be one of these intervals, and consider the difference $x(t;0;\eta_1) - x(t;0;\eta_2) = \varphi(t,\eta_1) - \varphi(t,\eta_2)$ which is almost periodic if there is an almost periodic solution at all. Since S' is invariant under ψ, $\varphi(m,\eta_1)$ and $\varphi(m,\eta_2)$ are endpoints of an interval in the complement of S'. This also uses the order preserving property of ψ. Since for $\epsilon > 0$ given there are only a finite number of intervals (a_n,b_n) in the complement of S' such that $|b_n - a_n| \geq \epsilon$, it follows that

$\lim_{m\to\infty}(\varphi(m,\eta_1) - \varphi(m,\eta_2)) = 0$. But by Lemma 9.6 we can find a $\delta > 0$ so that $|\varphi(m,\eta_1) - \varphi(m,\eta_2)| < \delta$ implies $|\varphi(t,\eta_1) - \varphi(t,\eta_2)| < \epsilon$ for $m \leq t \leq m+1$. Thus $\lim_{t\to\infty}(\varphi(t,\eta_1) - \varphi(t,\eta_2)) = 0$, which is impossible for a non-zero almost periodic function.

4. Ultimate boundedness. We have seen that stability of a solution is closely related to its almost periodicity. There is a notion, called ultimate boundedness which implies the existence of a periodic solution of a periodic system.

Definition 12.6. The solutions of $x' = f(x,t)$ are uniformly ultimately bounded if there exists an M so that for every $K > 0$ there is a $T(K)$ such that if $|x_o| \leq K$, then $|x_f(t;t_o;x_o)| \leq M$ if $t \geq t_o + T(K)$.

In some sense, the ball $|x| \leq M$ is exponentially stable. For periodic equations, this says that there is an integer m such that $|x_f(m\omega;0;x_o)| \leq M$ if $|x_o| \leq M$. The continuous map $x_o \to x_f(m\omega;0;x_o)$ then has a fixed point by the Brouwer fixed point theorem. Thus there is a solution with period $m\omega$.

We now present an example to show that for almost periodic systems, uniform ultimate boundedness does not imply the existence of an almost periodic solution.

Example 12.7. We now construct an example of an almost periodic equation with uniformly ultimately bounded solutions but no almost periodic solution. We take

$$y' = h(y,t) = \begin{cases} g(y,t) & ; \ |y| \le 3, \\ g(y,t) + \frac{|y|}{y} C(y^2 - 9) & ; \ |y| > 3, \end{cases} \tag{12.6}$$

where C is a constant such that for $M = \sup_{y,t} |g(y,t)|$ we have $M + 7C < 0$, where $g(x,t) = f(x+\rho t,t) - \rho$ and f is selected as in Example 12.5

Clearly $h(y,t)$ is continuous and almost periodic, we now show that all solutions of (12.6) are uniformly ultimately bounded. Notice that $h(4,t) \le M + 7C < 0$ and similarly $h(-4,t) > 0$ so that once a solution enters $|y| \le 4$ it never leaves as t increases. We claim that 4 is a uniform ultimate bound. Let $B > 4$ be given, and $y(t)$ a solution such that $|y(t_o)| < B$, say $4 < y(t_o) < B$. Then as long as $y(t) \ge 4$ we have

$$y'(t) \le M + C(y^2(t) - 9) = Cy^2(t) + (M - 9C)$$

so

$$y'(t)y^{-2}(t) \le C + (M-9C)y^{-2}(t) \le C + (M-9C)(16)^{-1} = D < 0.$$

Hence $(-y^{-1}(t))' \le D$ for $t \ge t_o$ as long as $y(t) \ge 4$. Integrating and rearranging we get for such t, that $y(t) \le [B^{-1} - D(t-t_o)]^{-1}$. Hence $|y(t)| \le 4$ for $t - t_o$ large uniformly for $0 < y(t_o) \le B$. A similar argument holds for $-B \le y(t_o) < 0$.

We now must show that no solution is almost periodic. Since y' is one sign in either of the regions $y > 4$ or $y < -4$, no almost periodic solution can be outside $|y| > 4$ at any point.

No almost periodic solution can satisfy $|y| \leq 3$ since it would then be an almost periodic solution to (12.5). Finally we now show that no solution which is ever say in $3 < y < 4$ can be almost periodic.

Let z now denote any solution of (12.6) such that $3 < z(t) < 4$ for some t. Let y be any solution of (12.5) such that $y(t_o) = z(t_o) > -3$. We claim that $y(t) \leq z(t)$ for $t \leq t_o$ and $y(t) \geq z(t)$ for $t \geq t_o$. We first show this for $y(t_o) > 3$. Notice that $(y-z)'(t_1) = -C[y^2(t_o) - 9] > 0$ whenever $(y-z)(t_1) = 0$ and $y(t_1) > 3$. Hence $y(t) > z(t)$ for $t > t_o$ and t near t_o. Since they cannot cross in $|y| \leq 3$ they cannot cross on $t > t_o$.($y(t_o) > 3$ implies $y(t) > -3$ for all t). A similar argument holds to the left. If $y(t_o) = 3$, then the result follows by continuity with respect to initial conditions since the above inequalities are strict. If $-3 < y(t_o) < 3$, then $y = z$ until $z(t) = 3$, then apply above. This implies in particular that no solution z of (12.6) enters both regions $y > 3$ and $y < -3$.

Let z be an almost periodic solution of (12.6). We show this leads to a contradiction. Since either $z(t) > 3$ or $z(t) < -3$ somewhere, we assume the former. Note that $z(0) > -3$. We pick out a solution $\bar{y}$ of (12.5) such that $\eta > \bar{y}(0)$ implies the solution of (12.5) starting at η crosses z or is always above z on $[0,\infty]$. In fact $\bar{y}(0) = \inf\{y(0) \mid y$ meets z on $[0,\infty)$, y a solution of (12.5)$\}$. By continuity with respect to initial conditions $\bar{y}(t) \leq z(t)$ on $[0,\infty)$. We will arrive at a contradiction by showing that $\bar{y}(t) \leq z(t) - \epsilon$ for some $\epsilon > 0$ and that this cannot happen.

To show that an $\epsilon > 0$ exists we argue in the following way. Let $\sup_{t \geq 0} z(t) = 3 + b$, $b > 0$. Let $\epsilon_1(\alpha) = \sup\{|g(y_1,t) - g(y_2,t)| : t \in R,\ 3 + \alpha \leq y_1, y_2 \leq 3 + b\}$ for $0 \leq \alpha \leq b$. Notice that as $\alpha \to b$, $\epsilon_1(\alpha) \to 0$ so that there exists an $\alpha \in (0,b)$ such that $\epsilon_1(\alpha) + C[6\alpha + \alpha^2] = d < 0$. Fix such an α. Since z is almost periodic and $\sup z > 3 + \alpha$, there exists a relatively dense set

$$E = \bigcup_{k=1}^{\infty} (t_k - \delta, t_k + \delta)$$

such that $t \in E$ implies that

$$z(t) \geq 3 + \frac{\alpha + b}{2},$$

Now if $0 < \eta < (b - \alpha)/2$, then $z(t) - \bar{y}(t) < \eta$ and $z(t) > 3 + \alpha$ imply that

$$(z - \bar{y})'(t) < \epsilon_1(\alpha) + C(z^2 - 9) < \epsilon_1(\alpha) + [6\alpha + \alpha^2]C = d < 0.$$

Consequently, if $z(t_k - \delta) - \bar{y}(t_k - \delta) < \eta$ then this inequality holds on $(t_k - \delta, t_k + \delta)$. Hence $(z - \bar{y})(t_k + \delta) \leq \eta + 2d\delta$ which is negative if η is small and thus for η small, $\bar{y}(t_k + \delta) > z(t_k + \delta)$ which can never happen. Thus $(z - \bar{y})(t_k - \delta)$ is bounded away from 0. This in fact implies that $(z - \bar{y})(t)$ is bounded away from 0 everywhere. For if not, let L be an inclusion interval for E. If $\eta > 0$ is given, suppose $(z - \bar{y})(t_o) < \eta e^{-ML}$ for some $t_o > 0$, then consider the solutions $\bar{y}$ and y_1 of (12.5) with $y_1(t_o) = z(t_o)$. Here M is the Lipschitz constant for $g(t,y)$. Then

$$(y_1 - \bar{y})(t) \le e^{M(t-t_o)} (y_1 - \bar{y})(t_o)$$

But since E has inclusion interval L, there exists a $t_k > t_o$ in E such that $(t_k - t_o) < L$. For this t_k we have $(y_1 - \bar{y})(t_k) < \eta$. But $\bar{y}(t_k) \le z(t_k) \le y_1(t_k)$. Since η was arbitrary we must have $\epsilon > 0$ such that $(z - \bar{y})(t) > \epsilon$ for all $t \ge 0$.

Now we show that this cannot happen. There are two cases. Either $\bar{y}(0)$ is in $S' - \bigcup_n(\{a_n\} \cup \{b_n\})$ or not. If $\bar{y}(0)$ is not, then there exists numbers η_1 and η_2 such that $\eta_1 \le \bar{y}(0) < \eta_2$, with η_1 and η_2 endpoints of an open interval in the complement of S'. But if we let $\varphi(t,\eta)$ be the solution of $x' = f(x,t)$ starting at η and $y(t,\eta)$ be the solution of (12.5) starting at η then

$$y(t,\eta_2) - y(t,\eta_1) = \varphi(t,\eta_2) - \varphi(t,\eta_1) \to 0$$

as $t \to \infty$. But $y(t,\eta_1) \le \bar{y}(t) \le z(t) \le y(t,\eta_2)$. Consequently $z(t) - \bar{y}(t) \to 0$ as $t \to +\infty$.

It $\bar{y}(0) \in S'$ and not an endpoint of the complement then we use a different argument. It is known that if η is not an endpoint, then $\varphi(t;\eta)$ is a recurrent motion, see Bhatia and

Szego [14] for example. This means, that if $\epsilon > 0$ is given, then there is a relatively dense set E_η so that for $t \in E_\eta$, $|\varphi(t,\eta) - \eta| < \epsilon$. By the uniform continuity of φ, we may assume that $E_\eta \supset \bigcup_{n=-\infty}^{\infty} (t_k - \delta, t_k + \delta)$ for some $\delta > 0$ and a relatively dense set of t_k. The importance of this remark, is that two such sets, which are clearly $\Delta - m$ sets, have a relatively dense intersection. We can conclude from this, that there is a $\delta > 0$ so that if η_1 and $\eta_2 \in S'$ but not endpoints and $|\eta_1 - \eta_2| < \delta$, then $|\varphi(t,\eta_1) - \varphi(t,\eta_2)| < \epsilon$ for a relatively dense set of t. Now take $\eta_1 = \eta$ not an endpoint and $\eta_2 = \bar{y}(0)$. Such an η_1 exists since $\bar{y}(0)$ is actually a condensation point of S' and there are only a countable number of endpoints. This implies, as before, that $|y(t,\eta) - \bar{y}(t)| < \epsilon$ on a relatively dense set and hence $\bar{y}(t) \leq z(t) \leq y(t,\eta)$ implies that $z(t_o) - \bar{y}(t_o) < \epsilon$ for some $t_o > 0$. This completes the contradiction and we conclude that z cannot be almost periodic.

5. Monotone $f(x,t)$. We have discussed several criteria which do not imply the existence of an almost periodic solution, we now turn to some positive results. We will look at the equation

$$x' = f(x,t) \tag{12.7}$$

where f is monotone in x, that is, if $x > y$ then $f(x,t) \leq f(y,t)$.

First note that if $f(x,t) \leq f(y,t)$ when $x < y$ then the change of variable $t \to -t$ replaces $f(x,t)$ by $-f(x,-t)$

so that (12.7) is satisfied. An example of such an equation might be $f(x,t) = f(x) + g(t)$ with $g \in AP(R)$ and f monotone, say $f(x) = -x$.

<u>Theorem 12.8</u>. Let $f(x,t)$ be a scalar function almost periodic in t uniformly for x in compact sets with f monotone decreasing in x, and such that all equations in the hull of (12.7) have unique solutions to initial value problems. If (12.7) has a bounded solution on $[0,\infty)$ then it has an almost periodic solution with module in the module of f.

<u>Proof</u>: We first get a solution φ that is bounded on R. Then we define $\lambda(\varphi) = \sup_{t\in R} \varphi - \inf_{t\in R} \varphi$. Note that λ is a subvariant functional since $\sup_{t\in R} T_\alpha\varphi(t) \leq \sup_{t\in R} \varphi(t)$ and $\inf_{t\in R} T_\alpha\varphi(t) \geq \inf_{t\in R} \varphi(t)$. By an argument similar to Theorem 5.1, there is a solution which minimizes λ. We now claim that if φ is a minimizing solution and $K = \text{Range}(\varphi)$ then the minimum solution must be unique in K. If $\lambda(\varphi) = \lambda(\psi)$, then we note that by unique solutions to initial value problems we may assume that $\varphi(0) < \psi(0)$. It follows that $\sup_{t\in R} \psi(t) \geq \sup_{t\in R} \varphi(t)$ and $\inf_{t\in R} \psi(t) \geq \inf_{t\in R} \varphi(t)$. But $u = \psi - \varphi$ satisfies $u'(t) = \psi'(t) - \varphi'(t) = f(\psi(t),t) - f(\varphi(t),t) \geq 0$ and $u(0) > 0$. Thus $\sup_{t\in R^+} \psi(t) \geq \sup_{t\in R^+} \varphi(t) + u(0)$. Hence ψ is not in the range of φ provided we can show that $\sup_{t\in R^+} \varphi(t) = \sup_{t\in R} \varphi(t)$. But this

is true, for if $\sup_{t\in R^+} \varphi < \sup_{t\in R} \varphi(t)$, take $\alpha_n' = n$ and $\alpha \subset \alpha'$ so that $T_\alpha f = f$ and $T_\alpha \varphi = \varphi_1$ exist uniformly on compact sets. Then $\sup_{t\in R} \varphi_1 \leq \sup_{t\in R^+} \varphi(t)$ and $\inf_{t\in R} \varphi_1(t) \geq \inf_{t\in R} \varphi(t)$ so that $\lambda(\varphi_1) < \lambda(\varphi)$. If $\psi(0) < \varphi(0)$ do the above argument on R^-. This proves that φ is the unique minimizing solution. Now if we take any equation in the hull, then there is a solution ψ, in K with $\lambda(\psi) = \lambda(\varphi)$ and the same argument as above gives uniqueness of the minimum. By Theorem 9.10, φ is almost periodic with module containment.

Another result that is true for scalar equation which we cannot prove for general ones is a theorem about uniformly stable solutions.

<u>Theorem 12.9</u>. Suppose the scalar equation $x' = f(x,t)$ is almost periodic in t uniformly for x in compact sets, have unique solutions of initial value problems for equations in the hull and has a bounded uniformly stable solution on $[0,\infty)$. Then there is a uniformly stable almost periodic solution on R with module in module of f.

<u>Proof</u>: We take $\alpha_n' \to \infty$ so that $g(x,t) = T_\alpha f(x,t)$ and $T_\alpha \varphi = \psi$ exist uniformly on compact sets for some $\alpha \subset \alpha'$. Then ψ is uniformly stable on R and is a solution of $x' = g(x,t)$. If further we pick α_n' so that $\varphi(\alpha_n') \to \sup_{t\in R^+} \varphi(t) = b$, then $\psi(0) = b$ so no other solution of $x' = g(x,t)$ lies above ψ and is in $[c,b]$ where $c = \inf_{t\in R} \psi$. Now we claim that

$\inf_{t\in R^-} \psi(t) = c$. If in fact, $d = \inf_{t\in R^-} \psi(t) > c$, we take

$\beta_n' \to -\infty$ so that $T_{\beta'}g(x,t) = g(x,t)$ and $\beta \subset \beta'$ so that $T_\beta\psi = \sigma$ exists uniformly on compact sets. Then σ is a solution of $x' = g(x,t)$ and $d \leq \sigma(t) \leq b$. Therefore $\sigma(t) \leq \psi(t)$ since no solution lies above ψ. Then

$c = \inf_{t\in R} \psi \geq \inf_{t\in R} \sigma \geq d > c$. This contradiction shows that $d = c$.

Next we claim that ψ is the only solution of $x' = g(x,t)$ in $[c,b]$. If $\chi(t)$ is another solution, then $\epsilon = \psi(0) - \chi(0) > 0$. Since $c \leq \chi(t) \leq \psi(t)$ on R^- we have a $t_o < 0$ so that $|\psi(t_o) - \chi(t_o)| < \delta(\epsilon/2)$ the δ of uniform stability. Then $|\psi(0) - \chi\ (0)| < \epsilon/2$ which is a contradiction. Thus ψ is the only uniformly stable solution of $x' = g(x,t)$ in $[c,b]$. By Theorem 11.15 we have our result.

6. Module containment. The conclusion in the previous two theorems that gives module containment of the almost periodic solution is not the result of the special circumstances of the hypothesis of the theorems. For scalar equations, almost periodic solutions always have module containment.

Theorem 12.10. Let the scalar almost periodic equation $x' = f(x,t)$ have unique solutions to initial value problems. If φ is an almost periodic solution, then $\mathrm{mod}(\varphi) \subset \mathrm{mod}(f)$.

Proof: Let $T_{\alpha'}f = f$ and pick $\alpha \subset \alpha'$ so that $\psi = T_\alpha\varphi$ exists uniformly. We need to show that $\psi = \varphi$. We suppose not, say

$\psi(t) < \varphi(t)$ for all t. Let ℓ be an inclusion length for $T(\varphi,\epsilon/3)$ where $\epsilon = \varphi(0) - \psi(0)$. Note that $T(\psi,\epsilon) = T(\varphi,\epsilon)$. Also for almost periodic functions it is easy to verify $\sup_{t\in R} \psi = \sup_{t\in R} T_\alpha\varphi = \sup_{t\in R} \varphi$. Thus there is a sequence t_n so that $\psi(t_n) - \varphi(t_n) \to 0$. By Lemma 9.6, there is an interval of length ℓ, say $[a,a+\ell]$ on which $|\psi(t) - \varphi(t)| < \epsilon/3$. This interval also contains a $\tau \in T(\varphi,\epsilon/3)$. Then $-\tau \in T(\varphi,\epsilon/3)$ also. We thus have $|\varphi(t-\tau) - \varphi(t)| < \epsilon/3$ and $|\psi(t-\tau) - \psi(t)| < \epsilon/3$ for all t. By the triangle inequality we have $|\varphi(t-\tau) - \psi(t-\tau)| < \epsilon$ for $t \in [a,a+\ell]$. This implies that $|\varphi(t) - \psi(t)| < \epsilon$ on $[a-\tau,a+\ell-\tau]$. Now $0 \in [a-\tau,a+\ell-\tau]$ so we have a contradiction. Thus $T_\alpha\varphi = \varphi$ and $\text{mod}(\varphi) \subset \text{mod}(f)$ by Theorem 4.5v.

7. Notes. Massera, The existence of periodic solutions of systems of differential equations, Duke Math. J. 17(1950), 457-475, has proved Theorem 12.1 and 12.2.

The results on equations on the torus through Theorem 12.3 are in Denjoý, Sur les courbes defines par les équations differentielles a le surface du tore, J. de Math. 11(1932), 333-375.

The example 12.5 is in Opial, Sur une équation differentielle presque-periodique sans solution presque-periodique, Bull. Acad. Polon. Sci Ser. Sci. Math. Astr. Phys. 9(1961), 673-676 and its extension to the results of article 4 are in Fink and Frederickson, Ultimate boundedness does not imply almost periodicity, J. Differential Equations, 9(1971), 280-283.

The results in Article 5 can be found in Opial, Sur les solutions presque-periodiques des équations differentielles du premier et du second ordre, Ann. Polonici. Math, 7(1959), 51-61

Theorem 12.9 was communicated to the author by Sell and Sacker, who proved it using dynamical systems.

Chapter 13

Second order equations

1. Introduction. We are going to study some special examples of second order equations for which one can get existence of almost periodic solutions. We will not include all the details since some would lead us far astray. On the other hand, second order equations are a rich source for further study so we will attempt to give the flavor of the ideas involved.

2. The Maximum Principle. Our first application is a generalization of Theorem 12.9.

Definition 13.1. Consider the equation

$$y'' = f(y,y',t). \tag{13.1}$$

We say that the equation satisfies the maximum principle if for arbitrary $a < b$ and solutions y and z, the inequalities $z(a) - y(a) \leq M$, $z(b) - y(b) \leq M$, and $0 \leq z(t) - y(t)$ on $[a,b]$ imply that $z(t) - y(t) \leq M$ on $[a,b]$.

Thus solutions that do not cross have the maximum of the difference at the endpoints.

Theorem 13.2. Let (13.1) satisfy the standard hypothesis on K and suppose that each equation in the hull satisfies the maximum principle. If there is a bounded solution on R, with a bounded derivative, then there is an almost periodic solution with module in the module of f.

Proof: We use the functional $\lambda(y) = \sup_{t\in k} y - \inf_{t\in k} y$. We have a minimizing solution as before. Then define K_0 to be a rectangle of the form $K_1 \times K_2$ where K_1 = range of a minimizing solution. The claim is that no other minimizing solution is in K_0. Say φ and ψ are two solutions in K_0. Let $u(t) = \varphi(t) - \psi(t)$ and suppose $u(t_0) = \epsilon > 0$ with $u'(t_0) > 0$. If $u(t_1) < u(t_0)$ for some $t_1 > t_0$, then one can find t_2 and t_3 such that $u(t_2) > u(t_3) = u(t_0)$ and $t_2 \in (t_0,t_3)$ and $u(t) > 0$ on $[t_0,t_3]$. This contradicts the maximum principle. Thus $u(t) \geq \epsilon$ on $[t_0,\infty)$. If $u(t_0) = \epsilon$ and $u'(t_0) < 0$, then as above, we have $u(t) \geq \epsilon$ on $(-\infty,t_0]$. Finally if $u'(t) \equiv 0$, then we get both. So in all cases $u(t) \geq \epsilon$ on a ray. Now as in Theorem 12.9 we argue that if φ is a minimizing solution, then $\sup_{t\in R^+} \varphi = \sup_{t\in R^-} \varphi$ so that $\sup_{t\in R} \varphi = \sup_{t\in R}(\psi + u) \geq \sup_{t\in R} \psi + \epsilon$ so not both φ and ψ can be in K_0. Now we appeal to Theorem 10.1.

It is of interest to have the conditions of the previous theorem in terms of the function f directly. This is possible.

Lemma 13.3. Suppose $f(y,z,t)$ is continuous and $f(y_1,z,t) < f(y_2,z,t)$ if $y_1 < y_2$. Then (13.1) satisfies the maximum principle. Conversely, if the maximum principle holds, then $f(y_1,z,t) \leq f(y_2,z,t)$ if $y_1 < y_2$.

Proof: Let $u(t) = \varphi(t) - \psi(t)$ for φ and ψ solutions of (13.1). If $u(t_0) \neq 0$ and $u'(t_0) = 0$, then

$u''(t_0) = f(\psi(t_0) + u(t_0),\ \psi'(t_0),t_0) - f(\psi(t_0),\ \psi'(t_0),t_0)$ so that

$u(t_0)\, u''(t_0) > 0$. Thus u can have no interior extremum on any interval. Conversely, let $y_0 > z_0$ and let y and z be the solutions to the initial value problems $y(t_0) = y_0$ $y'(t_0) = y_1$ and $z(t_0) = z_0$, $z'(t_0) = y_1$. These solutions exist on a small interval $[t_0 - \delta, t_0 + \delta]$ and $y(t) > z(t)$ there. By the maximum principle, $u(t) = y(t) - z(t)$ does not have an interior maximum. But $u'(t_0) = 0$ and $u(t_0) > 0$ then imply that $u''(t_0) \geq 0$. This is $f(y_0, y_1, t_0) \geq f(z_0, y_1, t_0)$. But y_1 and t_0 are arbitrary.

Another condition which implies the maximum principle is the above necessary condition and a Lipschitz condition in the second variable, see Jackson [15] for details. In any case, to apply Theorem 13.2 one needs a solution bounded on a ray. We will quote one such condition.

Example 13.4. Let $f(y,z,t)$ satisfy

(i) $f(a,0,t) \leq 0 \leq f(b,0,t)$ for all t and some $a < b$;

(ii) there are $c < 0 < d$ such that $f(x,c,t)$ and $f(x,d,t)$ do not change sign on $x \in [a,b]$ and $t \in R$;

(iii) f is almost periodic uniformly on $[a,b] \times [c,d]$; and

(iv) $g(y_1,z,t) > g(y_2,z,t)$ if $y_1 > y_2$ and $g \in H(f)$.

There is an almost periodic solution φ with $\operatorname{mod}(\varphi) \subset \operatorname{mod}(f)$ and $(\varphi, \varphi') \in [a,b] \times [c,d]$. The existence of a solution follows from Schmitt [16] and its almost periodicity from Theorem 13.2. Note that if $f(y,z,t) = h(y,z) + p(t)$ with $p \in AP(R)$, then the conditions on g above are quite simple and it is easy to give concrete examples.

There is a slight generalization of the maximum principle to vector systems of the form (13.1). Say the vector maximum principle holds if it holds in each component. Sufficient conditions analogous to Lemma 13.3 are given in Heimes [17]. We prove a theorem for such vector systems since it illustrates the principle of multiple functionals.

Theorem 13.5. Let (13.1) satisfy the vector maximum principle as an hypothesis on its hull and let $K_1 = \overset{n}{\underset{i=1}{X}} [a_i,b_i]$ and K_2 be compact sets. Let f satisfy the standard hypothesis on $K_1 \times K_2$. If there is a solution in $K_1 \times K_2$, then there is an almost periodic solution.

Proof: Define $\lambda_1(\varphi) = \sup_{t\in R} \varphi_1 - \inf_{t\in R} \varphi_1$ where φ_1 is the first component of a solution φ of (13.1) in $K_1 \times K_2$. Using the argument of Theorem 13.2 we get $[c_1,d_1] \subset [a_1,b_1]$ so that every solution in $S = [c_1,d_1] \times \overset{n}{\underset{i=2}{X}} [a_i,b_i]$ has the same first component. Then introduce $\lambda_2(\varphi) = \sup_{t\in R} \varphi_2 - \inf_{t\in R} \varphi_2$ on the solutions which are in S and get $[c_2,d_2] \subset [a_2,b_2]$ so that all solutions in $[c_1,d_1] \times [c_2,d_2] \overset{n}{\underset{i=3}{X}} [a_i,b_i]$ have the same first two components. A repitition of this argument gets a set on which there is exactly one solution and we finish by Theorem 10.1.

3. Uniqueness by Lypanov function. The crucial part of the arguments using functionals is to distinguish some solution. One way to do this is to get some set in which only one solution resides. In this section we will use a Lypanov function method to get unique solutions. We

illustrate the general results with several examples before quoting the general theorem.

Consider an equation of the form

$$x'' = f(x,x') + e(t) \tag{13.2}$$

with $e(t) \in AP(R)$ and x a scalar with f having continuous first partials. Note that conditions placed on this equation which depend only on the range of $e(t)$ are automatically satisfied by all equations in the hull.

<u>Lemma 13.6</u>. Assume that there is a compact rectangle K and a number λ so that

$$\Delta = \left(\frac{\partial f}{\partial y}(x,y) - \lambda\right)^2 - 4\left(\frac{\partial f}{\partial x}(x,y)\right) < 0 \tag{13.3}$$

for $(x,y) \in K$. Then (13.2) has at most one solution in K.

<u>Proof</u>: Let x_1 and x_2 be two solutions in K. Write $y = x_1 - x_2$ and $z = x_1' - x_2'$. Then $y' = z$ and $z' = \frac{\partial f}{\partial x}(u_t,v_t)y + \frac{\partial f}{\partial y}(u_t,v_t)z$ for some $(u_t,v_t) \in K$. Now consider $V(y,z) = y(z - \lambda y/2)$ with λ as defined in (13.3). Then $V'(t) = z^2 + (\frac{\partial f}{\partial y}(u_t,v_t) - \lambda)yz + \frac{\partial f}{\partial x}(u_t,v_t)y^2$ which is a positive definite quadratic form by (13.3). Thus V is a bounded nondecreasing function on R. Let $V(+\infty) = V_0 \neq 0$. Then there is a sequence $t_n \to +\infty$ so $V'(t_n) \to 0$. Since $z - \lambda y/2$ is bounded and $V_0 \neq 0$, $y(t_n) \not\to 0$. For an appropriate subsequence

$|y(t_n')| \geq \epsilon > 0$. But along this subsequence

$$V'(t_n') = (z + \frac{1}{2}\frac{\partial f}{\partial y}(u_t,v_t)y - y\lambda/2)^2 - \frac{\Delta}{4}y^2 \geq -M\epsilon^2/4 > 0, \quad \text{where}$$

$M = \max_K \Delta$. This contradiction shows that $V(+\infty) = 0$ so $V \equiv 0$ and $V' \equiv 0$. This gives $y \equiv z \equiv 0$.

To get specific examples we now appeal to theorems which give existence of bounded solutions.

Example 13.7. Let $f(x,y)$ be continuous and suppose

(i) $f(a,0) + e(t) \leq 0 \leq f(b,0) + e(t)$ for $a < b$, $t \in R$;

and

(ii) there are $c < 0 < d$ such that $f(x,c) + e(t)$ and $f(x,d) + e(t)$ do not change sign on $[a,b] \times R$;

or

(ii') there is a positive continuous function k so that $|f(x,y) + e(t)| \leq k(|y|)$ for $t \in R$, $x \in [a,b]$, $y \in R$ and $\int_0^\infty \frac{s\,ds}{k(s)} > b - a$.

It follows there is a $[c',d']$ so that there is a solution φ on R such that $\varphi(t) \in [a,b]$ and $\varphi'(t) \in [c',d']$. In the case of (ii) $c = c'$, $d = d'$. If further (13.1) is satisfied one gets an almost periodic solution.

As a specific example we may take $f(x,y)$ to be of an even simpler form.

Example 13.8. Consider

$$x'' = \alpha x + \beta x' + g(x) + e(t). \tag{13.4}$$

Let $a < b$ be such that for $t \in R$

$$\alpha a + g(a) + e(t) \leq \alpha b + g(b) + e(t)$$

and $g'(x) + \alpha > 0$ on $[a,b]$. Then (i) and (ii') are satisfied and so is (13.3) for $\lambda = \beta$. Thus there is an almost periodic solution to (13.4) under these conditions.

Example 13.9. Consider

$$x'' = m(x)x' + g(x) + e(t) \tag{13.5}$$

with $g(a) + e(t) \leq 0 \leq g(b) + e(t)$ for $a < b$. Let m and $g' > 0$ be continuous on $[a,b]$. Let $M(x) = \int_0^x m$ and

$$\varphi(u,v) = \begin{cases} \dfrac{M(u+v) - M(u)}{v} & \text{if} \quad v \neq 0, \\ m(u) & \text{if} \quad v = 0, \end{cases}$$

and

$$h(u,v) = \begin{cases} \dfrac{g(u+v) - g(u)}{v} & \text{if} \quad v \neq 0, \\ g'(u) & \text{if} \quad v = 0. \end{cases}$$

If further there is a λ so that $(\varphi(u,v) - \lambda)^2 - 4h(u,v) < 0$ when

$u \in [a,b]$ and $u+v \in [a,b]$ then there is an almost periodic solution to (13.5).

This does not quite fit into our previous framework. Instead of y and z as defined in Lemma 13.6 one defines $\xi = x' - M(x)$ and $\eta = g(x) + e(t)$ for each solution, and then $y = \xi_1 - \xi_2$ with $z = \eta_1 - \eta_2$. In this way one gets a unique solution since (i) and (ii') of Example 13.7 are satisfied.

The general framework for the above is now the following.

Definition 13.10. Let $V(x,t)$ be a function which satisfies

(i) $V: \Omega \times R \to R$ where Ω is open in E^n;

(ii) V is Lipschitz in x uniformly for $t \in R$;

(iii) $V(0,t) = 0$;

(iv) V is bounded on sets of the form $K \times R$, for K compact.

Then V is called a Lypanov function. We also define

$$\dot{V}_f(x,t) = \limsup_{h \to 0^+} \frac{V(x+hf(x,t),t+h) - V(x,t)}{h}$$

for f a continuous function such that $f(0,t) \equiv 0$.

Lemma 13.11. Let K be a compact set containing 0 and suppose there is a Lypanov function V and a scalar function $a(r)$ defined on R^+, so that $a(r) \geq 0$ and $a(r) = 0$ if and only if $r = 0$. If either $\dot{V}_f(x,t) \geq a(|x|)$ or $\dot{V}_f(x,t) \leq a(|x|)$ holds on $K \times R$, then $x \equiv 0$ is the only solution of $x' = f(x,t)$ in K.

Proof: We observe that if $w(t) = V(x(t),t)$ for x a solution of $x' = f(x,t)$, then for D^+ the right hand upper Dini derivate, $D^+w = \dot{V}_f(x(t),t) \geq a(|x|) \geq 0$. One argues now exactly as in Lemma 13.6. Since w is bounded, $w(\infty)$ exists so $D^+(t_n) \to 0$ as $t_n \to \infty$ and thus $|x(t_n)| \to 0$ and hence $V(x(t_n),t_n) \to 0$. This follows since $|V(x(t_n),t_n) - V(0,t_n)| \leq K|x(t_n)|$. Thus $w(\infty) = 0$. Similarly $w(-\infty) = 0$ so $V(x(t),t) \equiv 0$ and $\dot{V}_f(x(t),t) \equiv 0$ thus $a(|x(t)|) \equiv 0$ and it follows that $x \equiv 0$.

Lemma 13.12. If $f(x,t)$ is uniformly continuous and $f(x,t+\alpha_k) \to g(x,t)$ uniformly on compact sets, then the existence of a uniformly continuous Lypanov function satisfying $\dot{V}_f(x,t) \geq a(|x|)$ is inherited.

Proof: If $T_\alpha f = g$, then consider $V(x,t+\alpha_k)$. This family is uniformly bounded and uniformly equicontinuous so we may extract a subsequence that converges uniformly on compact sets, say to $W(x,t)$. The properties of a Lypanov function are easily shown to hold for W. What concerns us most here is the derivative condition. Fix $(x,t) \in K \times R$, and let $h > 0$ and small so that $x + \theta h\, f(x,t+\alpha_k) \in U$ for $0 \leq \theta \leq 1$ and all k large. Then

$$\frac{V(x + h\, f(x,t+\alpha_k),\, t + \alpha_k + h) - V(x,t+\alpha_k)}{h} \geq \dot{V}_f(x + \theta_k h\, f(x,t+\alpha_k), t + \alpha_k + h)$$

for a θ_k such that $0 \leq \theta_k \leq 1$ by use of the mean value theorem for Dini derivates. Bounding this by $a(|x + \theta_k h\, f(x,t+\alpha_k|)$ we take subsequences so that $\theta_k \to \theta_0$ and then let $k \to \infty$. We get

$$\frac{W(x + h\, g(x,t),\, t + h) - W(x,t)}{h} \geq a(|x + \theta_0 h\, g(x,t)|) .$$ Letting $h \to 0$

we get $\dot{W}_g(x,t) \geq a(|x|)$.

We now consider an equation of the form

$$x' = F(x,t) \tag{13.6}$$

and a bounded solution $\varphi(t)$. Then form $f(x,t) = F(x+\varphi(t),t) - F(\varphi(t),t)$. If $\varphi(t) \in K$, then let $S \equiv \{k_1 - k_2; k_i \in K\}$. This is a compact set which contains 0. Now look at the equation

$$x' = f(x,t). \tag{13.7}$$

This equation, now no longer almost periodic if F is, does have the property that $x \equiv 0$ is a solution and if this is the only solution in S, then the only solution of (13.6) in K is φ. Now if (13.7) has a Lypanov function satisfying Lemma 13.11 that is uniformly continuous, then not only is φ the unique solution of (13.6) in K, but $T_\alpha\varphi$ is the unique solution of $T_\alpha F$ in K since the equation (13.7) corresponding to $T_\alpha F$ has a Lypanov function. It then follows that φ is almost periodic.

Theorem 13.13. Let (13.6) satisfy the standard hypothesis and have a bounded solution so that (13.7) has a Lypanov function satisfying the hypotheses of Lemma 13.11 and 13.12. Then φ is almost periodic and $\operatorname{mod}(\varphi) \subset \operatorname{mod}(f)$.

Note that the condition (13.3) in Lemma 13.6 does imply that $\dot{V}(y,z,t) \geq c(y^2+z^2)$ for some c so that $a(r) = cr$ in this case.

As another example of this kind of argument consider the following.

Example 13.14. The equation

$$x'' + f(x)x' + g(x) = k\,p(t) \qquad (13.8)$$

where $p \in AP(R)$ and $k > 0$ and f and g are continuous is equivalent to

$$\begin{aligned} x' &= y - F(x) \\ y' &= -g(x) + k\,p(t) \end{aligned} \qquad (13.9)$$

if $F(x) = \int_0^x f(u)\,du$.

Assume that $\|p\| = 1$ and that one can find $a > b$ and $c < d$ so that $g(c) = -k$ and $g(d) = k$ with

$k < \min\{[F(d) - F(c)]f(x) + g(-a),\ [F(-b) - F(-a)]f(x) - g(d)\}$ on $[-a,d]$. Suppose also that $g(0) = 0$ and g' exists and satisfies $0 < g'(x) \leq \beta$ and that $f(x) \geq \alpha$ where $\beta < \alpha^2$. Note that the condition on k is satisfied if g is large enough at $\pm\infty$.

For a compact set K that contains a unique solution of (13.8) we propose the interior of the set whose boundary is the union of the following curves:

Γ_1: $y = F(x) + F(d) - F(c)$ for $-a \leq x \leq c$;

Γ_2: $y = F(d)$ for $c \leq x \leq d$;

Γ_3: $x = d$ for $F(d) - F(-b) + F(-a) \leq y \leq F(d)$;

Γ_4: $y = F(x) - F(-b) + F(-a)$ for $-b \leq x \leq d$;

Γ_5: $y = F(-a)$ for $-a \leq x \leq -b$;

Γ_6: $x = -a$ $F(-a) \leq y \leq F(-a) + F(d) - F(c)$.

It is an elementary exercise to show that no trajectory starting in K can exit through any of the segments Γ_i. Thus there is a solution in K on a ray of the form $[t_0,\infty)$. By the almost periodicity there is one on R. The above argument depends only on the estimate $\|p\| = 1$. Thus every equation in the hull has the same properties as (13.8) with respect to K. In fact we can show that (13.8) has precisely one solution in K. Let V be defined in the following way. If (x,y) is a solution in K, suppose $(\bar{x},\bar{y})$ is another and look at $(u,v) = (x,y) - (\bar{x},\bar{y})$. Then $u' = v - m(t)u$ and $v' = -h(t)u$ where

$$m(t) = \begin{cases} \dfrac{F(\bar{x}(t)+u(t)) - F(\bar{x}(t))}{u(t)} & \text{if} \quad u(t) \neq 0 \\ f(\bar{x}(t)) & \text{if} \quad u(t) = 0 \end{cases}$$

and

$$h(t) = \begin{cases} \dfrac{g(\bar{x}(t)+u(t)) - g(\bar{x}(t))}{u(t)} & \text{if} \quad u(t) \neq 0 \\ g'(\bar{x}(t)) & \text{if} \quad u(t) = 0. \end{cases}$$

Then $\alpha \leq m(t) \leq M$ where M is an upper bound for f on K. Now define $V = v^2 + (v - zu)^2$ where z is a solution of $z' = z(m-z)$

satisfying $\alpha \leq z(t) \leq M$. Such a solution exists since if $z(t_0) \in (\alpha, M)$ then $z(t)$ satisfies the same inequality to the right since $\alpha \leq m(t) \leq M$. Taking a sequence of t_0 going to $-\infty$ gives a sequence of solutions whose limit along some subsequence gives the required z. With V so defined, it is easy to see that

$$\dot{V} = -2z[\{v - (z^2-h)\frac{u}{z}\}^2 + \frac{u^2}{z^2} h(z^2-h)] \geq -2M[\{v - (z^2-h)\frac{u}{z}\}^2 + \frac{u^2}{z^2} h(z^2-h)].$$

Since $z^2-h \geq \alpha^2-\beta > 0$. Thus $\dot{V} \geq -2M[v^2 + (z^2-h)\frac{u^2}{z^2} + \frac{u^2}{z^2} h(z^2-h)]$

$\geq -2M[v^2 + u^2 M_1]$ where M_1 is an upper bound for $\frac{z^2 - h^2}{z^2}(1+h)$ on K. Thus $a(r) = 2M \max(1, M_1) r$ is a function that satisfies Lemma 13.12, and the uniqueness of solutions is verified. Thus (13.8) has an almost periodic solution.

Note that the existence of certain stable solutions gives the existence of Lypanov functions by so-called converse theorems. See Yoshizawa's book for examples.

4. Uniqueness by comparison. The idea to be exposed in this section is a generalization of the maximum principle. If $f(y_1,z,t) > f(y_2,z,t)$ for $y_1 > y_2$, let $m(t)$ be a function that gives this inequality, that is, suppose $f(y_1,z,t) - f(y_2,z,t) \geq m(t)(y_1 - y_2)$. If further $|f(y,z_1,t) - f(y,z_2,t)| \leq K|z_1 - z_2|$, then let $u(t) = x(t) - y(t)$ for x and y two solutions of (13.1). Then $u'' \geq u(t)m(t) - K|u'(t)|$. Suppose we have conditions on m which imply that this differential inequality has unbounded solutions. Then (13.1) can have at most one

bounded solution, the situation we want. If we assume that $m \geq 0$ then we have the maximum principle. But if m is now allowed to be negative, we may get a new result. Thus we begin a study of second order linear almost periodic equations. Consider

$$Ly = y'' + p(t)y' + q(t)y = 0, \qquad (13.10)$$

with p and q in $AP(R)$.

<u>Lemma 13.15</u>. If y is a nontrivial solution of (13.10) then either y has infinitely many zeros or at most one.

<u>Proof</u>: Suppose $\bar{y}$ is a non-trivial solution with two zeros, say at $a < b$ and such that $\bar{y} > 0$ on (a,b). We may assume that $\bar{y}(\frac{a+b}{2}) = 1$ and $\bar{y}(b+\delta) < 0$ for some $\delta > 0$. Since $\|p\|$ and $\|q\|$ are finite, there is an $\epsilon > 0$ so that if $\|p_1 - p\| < \epsilon$ and $\|q_1 - q\| < \epsilon$, and z is a solution of $y'' + p_1 y' + q_1 y = 0$ with $z(t_0) = y(t_0)$, $z'(t_0) = y'(t_0)$ for y a solution of (13.10), then $|y(t) - z(t)| < \min\{1, -y(b+\delta)\}$ on $[t_0, t_0 + b - a + \delta]$. Actually, ϵ is independent of t_0. Now, if n is a given integer, pick $\tau_n \in T(p,\epsilon) \cap T(q,\epsilon) \cap [n,\infty)$. Consider the initial value problem $z'' + p(t+\tau_n)z' + q(t+\tau_n)z = 0$, $z'(a-\tau_n) = \bar{y}'(a)$, $z(a-\tau_n) = 0$. Then $\bar{y}(t) = z(t-\tau_n)$ since $\bar{y}$ satisfies the correct initial value problem. Thus $z(b-\tau_n) = 0$, and $z(b+\delta-\tau_n) = y(b+\delta)$. By choice of τ_n and ϵ the function y^* defined by $y'' + py' + qy = 0$, $y(a-\tau_n) = 0$, $y'(a-\tau_n) = \bar{y}'(a)$ is within $\min\{1, -\bar{y}(b+\delta)\}$ of z, hence is positive at $\frac{a+b}{2} - \tau_n$ and negative at $b+\delta$. Thus y^* has a zero on that

interval. By the Sturm oscillation theorem so does $\bar{y}$. Since n was arbitrary, $\bar{y}$ has infinitely many zeros.

A corollary to the proof is that every equation in the hull has the same character.

Lemma 13.16. Let p and $q \in AP(R)$ and suppose (13.10) has no non-trivial solution with more than one zero. Then the same is true for every equation in the hull.

Proof: By the Sturm oscillation theorem either all or no solutions have at most one zero. If some equation has a solution with three zeroes, approximate any other equation in the hull on the fixed interval as in the proof of Lemma 13.15 to get at least two zeros, hence infinitely many.

Lemma 13.17. If (13.10) has no solution with more than one zero and $\int_0^\infty e^{-\int_a^x p(t)\,dt}\,dx = +\infty$, then (13.10) has an unbounded solution on $[t_0,\infty)$. In fact every solution with a zero is unbounded.

Proof: Let u be a solution which is positive on $[t_0,\infty)$. Write

$$P(t) = \exp(-\int_0^x p(t)\,dt). \tag{13.11}$$

Then $v(x) = u(x)\int_{t_0}^x \frac{P(t)}{u^2(t)}\,dt$ is another solution that is positive on (t_0,∞) and independent of u. Then $\left(\frac{u}{v}\right)' = -\frac{P(t)}{v^2(t)}$ on (t_0,∞).

If v is bounded, then say $v^2(t) \le c^2$. We have

$\frac{u}{v}(t) - \frac{u}{v}(t_0+1) \le -c^{-2}\int_{t_0+1}^{t} P(s)ds$ for $t \ge t_0+1$. As $t \to +\infty$ we get $\frac{u}{v}(t) \to -\infty$ which is impossible. Thus v is unbounded.

We now extend this lemma to differential inequalities.

Lemma 13.18. Suppose (13.10) has no solutions with two zeros and $\int^{\infty} P = +\infty$ where P is defined by (13.11). If v is twice differentiable and $Lv \ge 0$, whenever $v \ge 0$, $v(a) = 0$, $v(x) > 0$ on $(a,a+\delta)$. Then $v \ge 0$ on (a,∞) and $\lim_{t\to+\infty} v(t) = +\infty$.

Proof: Suppose first that $Lv = 0$. Then $v > 0$ on (a,∞). If $Lu = 0$ and $u(a-\epsilon) = 0$ $u'(a-\epsilon) = 1$, then $u > 0$ on $(a-\epsilon,\infty)$ and $v(t) = u(t)\int_a^t \frac{P(s)}{u^2(s)}ds$ and by the previous lemma v is unbounded.

If $Lv \ge 0$ and $v'(a) > 0$, take u so that $Lu = 0$ $u(a) = 0$, $u'(a) = v'(a) = 0$. Write $W = u'v - v'u$ so that $W(a) = 0$ and $W' + pW \le 0$. Thus $\left(\frac{v}{u}\right)' = \frac{-W}{u^2} \ge 0$ so $\frac{v}{u}$ is an increasing function and $v(t) \ge u(t)$ on $[a,\infty)$. By the above, u is unbounded so v is also. If $Lv \ge 0$ and $v'(a) = 0$, then we take $u'(a) = 1$ and the same W. Note that $W(a) = 0$ and so $W \le 0$ and $\left(\frac{v}{u}\right)' \ge 0$. Since v and u are both positive at $(a+\delta)$ for small δ we have $v(x) \ge u(x)\frac{v(a+\delta)}{u(a+\delta)}$ on $(a+\delta,\infty)$ since v continues to be positive.

We now can do rigorously, the outline at the beginning of this section.

Theorem 13.9. Let $f(y, z, t)$ be almost periodic in t uniformly for (y, z) in compact sets. Suppose there is a bounded solution φ to (13.1) say in K. If $f(y_1, z_1, t) - f(y_2, z_2, t) \geq -p(t)(z_1 - z_2) - q(t)(y_1 - y_2)$ when $(y_i, z_i) \in K$ and $y_1 > y_2$, with p and $q \in AP(R)$, and $Ly = y'' + py' + qy$ has no solutions with more than one zero and $\int_0^\infty P = \int_{-\infty}^0 P = +\infty$, then φ is almost periodic and $\mathrm{mod}(\varphi) \subset \mathrm{mod}(f)$.

Proof: As before we get a solution on all of R in K. If there are two solutions, way φ and ψ, then set $v = \varphi - \psi$. If $v = 0$ somewhere, we may choose the notation so that $v(t_0) = 0$ and $v(t) \geq 0$ on $[t_0, t_0 + \delta]$. On that interval $Lv \geq 0$ so we may apply Lemma 13.8 to get v unbounded on R^+ which is a contradiction. If v is never zero, we may choose the notation so that $v > 0$ on all R and hence $Lv > 0$ on all of R. If $v'(a) > 0$ for some a, take $u(a) = v(a)$, $u'(a) = v'(a)$, and $Lu = 0$. If $w = v - u$ then $Lw > 0$ and $w(a) = a'(a) = 0$. Since $w''(a) = Lw(a) > 0$, $w > 0$ near a. By Lemma 13.18, and its analogue on $(-\infty, t_0]$, $w(t) \to \infty$ as $t \to \pm\infty$. Thus $v \geq u$ and the difference goes to $+\infty$. We need only show that u is bounded below. In fact u is unbounded above. For if u has a zero to the left of a, then it is unbounded by Lemma 13.17. If u has a zero to the right of a, then u is unbounded on the left, and we appeal to the analogue of Lemma 13.17 using the hypothesis that $\int_{-\infty}^{0} P = +\infty$. In either case v is unbounded. The case when $v > 0$ and $v'(a) < 0$ is handled similarly. If $v > 0$ and $v' \equiv 0$, then $v \equiv c > 0$ so we use an isolation argument that the bounded solution is unique in its range. Now we can appeal to a variety of theorems to complete the proof, Theorem 10.1 for example.

There are other interesting questions that arise out of the study of the disconjugacy of second order linear differential equations with almost periodic coefficients. In fact, the Riccati equation gotten by making the change of variable $z = y'y^{-1}$ in (13.10) is a good candidate for finding an almost periodic differential equation with a unique bounded solution that is not almost periodic. That is, if this is possible.

On the other hand, the equations we have been studying in this section have the opposite monotonicity in f that the Lienard equations to be studied in the next section.

5. A forced Lienard equation. One of the most studied second order equation is the Lienard type

$$x'' + f(x)x' + g(x) = p(t) . \tag{13.12}$$

The most definitive result in this direction is when $g(x) = x$. The most common way to handle (13.12) is to use the two dimensional system where

$$\begin{aligned} x' &= y - F(x) \\ y' &= -g(x) + p(t) \\ F(x) &= \int_0^x f(s)\,ds. \end{aligned} \tag{13.13}$$

<u>Theorem 13.10</u>. Suppose $f > 0$ and is continuous except at isolated points. Let $g(x) = x$. Then (13.12) has an almost periodic solution if and only if

$$F(\infty) - F(-\infty) > \pi\, M(p(t)e^{it}).$$

We consider this theorem as a special case of

$$\begin{aligned} x' &= y - G_1(x) + h_1(t) \\ y' &= -x - G_2(y) + h_2(t) \end{aligned} \tag{13.14}$$

where the h_i are in $AP(R)$ and G_i are continuous and non-decreasing. Let $\chi(s) = M_t\{h_1(t)\cos(t-s) - h_2(t)\sin(t-s)\}$ and $\beta = \max_s M(s)$.

<u>Theorem 13.11</u>. Suppose that (13.14) satisfies the above hypothesis. Then (13.14) has an almost periodic solution if and only if

$$G_1(\infty) - G_1(-\infty) + G_2(\infty) - G_2(-\infty) > \pi\beta \tag{13.15}$$

To prove this theorem we prove a sequence of Lemmas.

Lemma 13.12. If G_i are continuous and $\sup G_1 - \inf G_1 + \sup G_2 - \inf G_2 \leq \pi\beta$, then (13.14) has an almost periodic solution only if $\beta = 0$.

Proof: Note that we are not assuming the hypothesis of the theorem here. Let $\sin^+ t = \max(\sin t, 0)$ and $\sin^-(t) = \min(\sin t, 0)$, and $S(t) = (\sup w)\sin^+(t-T) - (\inf w)\sin^-(t-T)$ for w some almost periodic function. Then $S(t) - w(t)\sin(t-T) \geq 0$ so $M(S) - M(w(t)\sin(t-T)) \geq 0$ with equality if and only if $S(t) \equiv w(t)\sin(t-T)$. This gives $M(w(t)\sin(t-T)) \leq (\sup w - \inf w)\frac{1}{\pi}$ with equality only if $w(t)$ is a constant. Now if (13.14) has an almost periodic solution (x,y) with $\beta > 0$, then for some real T, $\beta = \chi(T)$. We then compute $M(G_1(x(t))\cos(t-T) - G_2(y(t))\sin(t-T))$ $= M(y(t)\cos(t-T) + h_1(t)\cos(t-T) - x'(t)\cos(t-T) + y'(t)\sin(t-T)$ $+ x(t)\sin(t-T) - h_2(t)\sin(t-T)) = \beta + M((y-x')\cos(t-T) + (y'+x)\sin(t-T))$. But integrating the x' and the y' terms by parts this last mean value is 0. Now taking $w = G_1(x(t))$ and $G_2(y(t))$ in turn we have (using an appropriate translation of axes) $\beta = M(G_1(x(t))\cos(t-T) - G_2(y(t))\sin(t-T)) \leq \frac{1}{\pi}(\sup G_1(x) - \inf G_1(x) - \sup G_2(y) - \inf G_2(y)$ with equality if and only if G_1 is constant on range of x and G_2 constant on range of y. If these are constant, then $\beta = 0$, if they are not constant, we get $\beta < \beta$ by the hypothesis of the lemma. This is a contradiction.

We thus see that the hypothesis of the theorem show that $\beta > 0$ in Lemma 13.12 so the necessity of (13.15) is proved.

We thus turn to the sufficiency.

Lemma 13.13. Let the hypothesis of the Theorem 13.11 be satisfied. Then every equation in the hull of (13.14) has at most one bounded solution or there is an almost periodic solution.

Proof: Let $d^2 = (x_1 - x_2)^2 + (y_1 - y_2)^2$ for (x_i, y_i) bounded solutions of the same equation in the hull. Then $dd' = -(x_1 - x_2)[G_1(x_1) - G_1(x_2)] - (y_1 - y_2)[G_2(y_1) - G_2(y_2)] \leq 0$ since the G_i are non-decreasing. Thus any bounded solution is uniformly stable. But if d^2 is finite at $-\infty$ say $= \delta^2$, take $\alpha'_n = -n$ and $\alpha \subset \alpha'$ so that (13.14) is transformed into itself. We get $(x_i, y_i) \to (\bar{x}_i, \bar{y}_i)$ a solution to (13.14) and $\bar{d}^2 = (\bar{x}_1 - \bar{x}_2)^2 + (\bar{y}_1 - \bar{y}_2)^2 \equiv \delta^2$ and hence $(\bar{x}_1 - \bar{x}_2)(G_1(\bar{x}_1) - G_1(\bar{x}_2)) + (\bar{y}_1 - \bar{y}_2)(G_2(\bar{y}_1) - G_2(\bar{y}_2)) \equiv 0$. This can happen only if $G_1(\bar{x}_1) \equiv G_1(\bar{x}_2)$ and $G_2(\bar{y}_1) \equiv G_2(\bar{y}_2)$. But if $u = \bar{x}_1 - \bar{x}_2$ and $v = \bar{y}_1 - \bar{y}_2$, then $u' = v$, $v' = -u$, and $u^2 + v^2 \equiv \delta^2$. Thus $u(t) = \delta \sin(t-T)$ and $v(t) = \delta \cos(t-T)$. Consequently $\bar{x}_1 = \bar{x}_2$ only at isolated points. Since G_1 is non-decreasing, this implies $G_1(\bar{x}_1) \equiv c_1$ and similarly $G_2(\bar{y}_1) = c_2$. Now $\bar{x}_1$ and $\bar{y}_1$ are bounded solutions of

$$\bar{x}_1' = \bar{y}_1 - c_1 + h_1(t)$$
$$\bar{y}_1' = -\bar{x}_1 - c_2 + h_2(t)$$

which is a linear system with constant linear part. Thus $\bar{x}_1$ and $\bar{y}_1$ are almost periodic.

We note that the proof of Theorem 13.11 will be completed if we show the existence of a bounded solution. This follows since, either every equation in the hull has a unique bounded solution or some equation has an almost periodic one. Note also that we need only show the existence of a solution bounded in the future.

To do the existence of a bounded solution, we switch to polar coordinates by $x = r\cos\varphi \quad y = r\sin\varphi$ to get the system

$$\begin{aligned} r' &= \cos\varphi[h_1 - G_1(r\cos\varphi)] + \sin\varphi[h_2 - G_2(r\sin\varphi)] \\ r(1+\varphi') &= \cos\varphi[h_2 - G_2(r\sin\varphi)] - \sin\varphi[h_1 - G_1(r\cos\varphi)] \end{aligned} \tag{13.16}$$

The basic idea is to find quadrants in which $r' \ll 0$ and $\varphi' > -1$ or in which $\varphi' < -\frac{1}{2}$ and $r' \leq B$. Here $r' \ll 0$ means that $r' \to -\infty$ as $r \to \infty$. Thus the possible gain in r as φ traverses the quadrants of the second kind is retrieved where $r' \ll 0$ or else r remains bounded. There are many cases to consider. We do only two. For example, if both G_1 and G_2 are bounded then r' and $r(1+\varphi')$ are bounded for all t. So pick $\epsilon > 0$ so that $G_1(\infty) - G_1(-\infty) + G_2(\infty) - G_2(-\infty) > \pi(\beta+4\epsilon)$ and choose T^* so that $T > T^*$ implies that $\frac{1}{T}\int_0^T [\cos(\theta-t)h_1(t) + \sin(\theta-t)h_2(t)]dt < \beta + \epsilon$. Let $T^* \leq T \leq 2T^*$. Then $|r(t_0+t) - r(t_0)| \leq B$ if $0 \leq t \leq T$ for some B independent of t_0. If $R = r(t_0) > 2B$ then $\frac{1}{r(t_0+t)} \leq (r(t_0) - B)^{-1}$. Since $|r(1+\varphi')|$ is bounded say by C, we now have $|1+\varphi'| \leq C(R-B)^{-1}$ so that $|\varphi(t+t_0) - \varphi(t_0) + t| < T\,C(R-B^{-1})$ if $0 \leq t \leq T$. Thus

$$\varphi(t+t_0) = \varphi(t_0) - t + \frac{D(t)}{R} \tag{13.17}$$

where D is a bounded function which is independent of R on $[0,T]$, $(R \geq 2B)$. Similarly $r(t) = r(t_0) + B(t)$, B bounded.

To get other estimates we state a Lemma.

<u>Lemma 13.14.</u> If F is bounded and nondecreasing and $\epsilon > 0$ is given, then there is a T^* and R_1 such that if $T > T^*$ and $r > R_1$, then $\frac{1}{T}\int_0^T \cos(\theta-t)F(r\cos(\theta-t))dt > \frac{F(\infty) - F(-\infty)}{\pi} - \epsilon$.

<u>Proof</u>: For $2n\pi < T < 2(n+1)\pi$ we have

$\frac{1}{T}\int_0^T \cos(\theta-t)\ F(r\cos(\theta-t)dt = \frac{n}{T}\int_0^{2\pi} \cos(\theta-t)\ F(r\cos\theta-t)dt$

$+ \frac{1}{T}\int_{2\pi n}^T \cos(\theta-t)\ F(r\cos\theta-t)dt$. First pick $T > T^*$ so that the last integral is at most $\epsilon/2$ in modulus. This is independent of r. Now as $r \to +\infty$ in the first integral, we use the Lebesgue dominated convergence theorem to get $\int_0^{2\pi} \cos(\theta-t)\ F(r\cos(\theta-t))dt \to \int_0^{2\pi} \cos(\theta-t)\ G(t)dt$ where $G(t) = F(\infty)$ if $\cos(\theta-t) > 0$ and $G(t) = F(-\infty)$ if $\cos(\theta-t) < 0$. Now $\int_0^{2\pi} \cos(\theta-t)\ G(t)dt = [2\ F(\infty) - 2F(-\infty)]\int_0^{\pi/2} \cos s\ ds$ $= 2F(\infty) - 2F(-\infty)$. Pick T^* so that $T > T^*$ implies that $\frac{2n\pi}{T}$ is near 1 and then R_1 so that $r > R_1$ implies

$\frac{1}{2\pi}\int_0^{2\pi} \cos(\theta-t)\ F(r\cos(\theta-t))dt = \frac{2F(\infty) - 2F(-\infty)}{2\pi} \pm \epsilon/2$, and we are done.

We can now estimate the integral (for $T > T^*$ from Lemma 13.14) $A_1 = \frac{1}{T}\int_0^T \cos[\varphi(t+t_0)]\ G_1(r(t_0+t)\cos\varphi(t+t_0))dt$ if

$R = r(t_0) > R_1 - B \geq 2B$ and sufficiently large.

$$A_1 = \frac{1}{T}\int_0^T \cos[\varphi(t_0)-t+\frac{C(t)}{R}]G_1((r(t_0)+B(t))\cos[\varphi(t_0)-t+\frac{C(t)}{R}])dt$$

$$> \frac{1}{T}\int_0^T \cos(\varphi(t_0)-t)G_1((r(t_0)+B(t))\cos(\varphi(t_0)-t))dt - \epsilon/2 \quad \text{if } R \text{ is}$$

large. Note $|B(t)| \leq B$. Note that R large here is independent of T. Now use lemma 13.14 to get $A_1 > \frac{1}{T}\ \frac{G_1(\infty) - G_1(-\infty)}{\pi} - \epsilon$. Similarly,

$$A_2 = \frac{1}{T}\int_0^T \sin\,\varphi(t+t_0)G_2(r(t+t_0)\sin\,\varphi(t_0+t))dt > \frac{1}{T}\ \frac{G_2(\infty) - G_2(-\infty)}{\pi} - \epsilon.$$

Finally $A_3 = \frac{1}{T}\int_0^T \cos(\varphi(t_0+t))h_1(t) + \sin(\varphi(t_0+t))h_2(t)dt$

$= \frac{1}{T}\int_0^T \cos(\varphi(t_0)-t+\frac{C(t)}{R})h_1(t) + \sin(\varphi(t_0)-t+\frac{C(t)}{R})h_2(t)dt$ converges to

$\frac{1}{T}\int_0^T [\cos(\varphi(t_0)-t)h_1(t) + \sin(\varphi(t_0)-t)h_2(t)]dt$ as $R \to \infty$ uniformly in T on $(T^*,2T^*)$, so is less than $\beta + 2\epsilon/\pi$ for $R \geq R_2$.

So having picked T^* as in Lemma 13.14 we pick R large so our three estimates hold. Then $r(t_0+T) - r(t_0) = -TA_1 - TA_2 + TA_3$

$$< T(\beta+2\tfrac{\epsilon}{\pi}) - \frac{T}{\pi}[G_2(\infty) - G_2(-\infty) - \epsilon] - \frac{T}{\pi}[G_1(\infty) - G_1(-\infty) - \epsilon]$$

$$= \frac{T}{\pi}[\beta\pi - (G_2(\infty) - G_2(-\infty) + G_1(\infty) - G_1(\infty))] + T(\frac{4\epsilon}{\pi}) < \frac{T}{\pi}(-4\epsilon) + T(\frac{4\epsilon}{\pi}) = 0.$$

Thus $r(t_0+T) < r(t_0)$ and so we iterate to get $r(nT) < r(T)$ or $r(t) \leq R^*$ for all t. Since r' is bounded it follows that r is bounded on $[T,\infty)$ as required.

When one of G_1 or G_2 is unbounded, one observes that $\cos\,\varphi\,G_1(r\cos\varphi)$ and $\sin\,\varphi\,G_2(r\sin\varphi)$ are bounded below, in fact by $-|G_1(0)|$ and $-|G_2(0)|$ respectively. Thus r' is bounded above by $\|h_1\| + \|h_2\| + |G_1(0) + |G_2(0)| = C$. As an example, if G_1 is unbounded above and G_2 is bounded, we look at a sector where $\sin \varphi > 0$ and $\cos \varphi > 0$, there $\cos \varphi\, G_2(r \sin \varphi) \to +\infty$ as $r \to \infty$. Taking this sector to be $(\frac{\pi}{4}, \frac{3T}{4})$, we look at r' there. If $r > R$ and R is large enough then $|\frac{d\varphi}{dr}| < \epsilon$ in this sector. Also $r' < 0$ there. Thus r decreases as the solution passes through this sector by as much as one likes requiring only that r being large when it enters. On the other hand if $r' \geq 0$ and r is large the solution enters the sector before r has increased by $4\pi C$. Thus the solution is bounded to the right. If both G_1 and G_2 are unbounded merely observe that $r' < 0$ for all large r. Then cases can be handled similarly.

<u>Question M</u>. Handle the above situation when $g(x)$ is not x.

<u>6. Other Lienard equations.</u> We briefly give some other conditions that give almost periodic solutions to equations of the type (13.12). Consider

$$x'' + k f(x)x' + g(x) = k p(t) \tag{13.18}$$

with $k > 0$ and $p \in AP(R)$. Define $F(x) = \int_0^x f(s)ds$, $G(x) = \int_0^x g(s)ds$ and $P(t) = \int_0^t p(s)ds$, with the assumptions that $f > 0$, $F(\pm\infty) = \pm\infty$, $x\,g(x) > 0$ for $x \neq 0$ with $G(\pm\infty) = +\infty$, and P bounded. Reuter [18 has shown that solutions of (13.18) are ultimately bounded with the bounds $|x(t)| \leq x_0$ and $|x'(t)| \leq x_0'$ independent of k. If further, $g' > 0$ and g'' is bounded on $[-x_0, x_0]$, then there is a k_0 so that if $k \geq k_0$, (13.18) has an almost periodic solution. If $g(x) = x$, then $k_0 = 0$.

On the other hand Opial [357] has shown that if $\|p\| = k$ and there are constants $a < c < 0 < d < b$ such that $g(d) = k$, $g(c) = -k$, $k < [F(b) - F(d)]f(x) + g(a)$, $k < [F(c) - F(a)]f(x) - g(b)$ for $x \in [a,b]$, $f(x) \geq m > 0$, $0 < g' \leq \ell < \lambda_0 m^2$ on $[a,b]$ with g strictly increasing and λ_0 a solution of $\ln \lambda = [\pi + 2\cot^{-1}\sqrt{4\lambda - 1}]\,(4\lambda - 1)^{-1/2}$, then $x'' + f(x)x' + g(x) = p(t)$ has an almost periodic solution.

7. <u>An application of column dominance</u>. Another way to look at equation (13.12) or (13.18) with $k = 1$, is to consider it as quasi-linear in the following manner. Write $u = \lambda_1 x$ and $v = \lambda_2(x' + F(x))$ where $F(x) = \int_0^x f(s)\,ds - bx$ and the positive numbers λ_1, λ_2 and b are to be chosen. Then (13.12) becomes

$$\begin{aligned} u' &= -\lambda_1 F(\lambda_1^{-1}u) + \lambda_1\lambda_2^{-1} v \\ v' &= b\lambda_2 F(\lambda_1^{-1}u) - \lambda_2 g(\lambda_1^{-1}v) - bv + \lambda_2 p(t) \end{aligned} \tag{13.19}$$

We will assume that $g(0) = 0$ and $g'(0)$ exists. So define $k(x) = x^{-1}g(x)$ and $h(x) = x^{-1}\int_0^x f(s)\,ds$. Let

$$A(u) = \begin{pmatrix} -h(\lambda_1^{-1}u) + b & \lambda \\ \lambda^{-1}[\,h(\lambda_1^{-1}u) - b^2 - k(\lambda_1^{-1}u)\,] & -b \end{pmatrix}$$

with $\lambda = \lambda_1\lambda_2^{-1}$. Then (13.19) becomes

$$\begin{pmatrix}u\\v\end{pmatrix}' = A(u)\begin{pmatrix}u\\v\end{pmatrix} + \begin{pmatrix}0\\\lambda_2 p(t)\end{pmatrix} \qquad (13.20)$$

We want to let u and v be a local solution to this equation and find conditions under which $A(u)$ is column dominant. If there is a set $|u| \le A^*$ and $|v| \le B^*$ on which $A(u)$ is column dominant, then u, and v starting there remain there and so we get a bounded solution. Next if we want some stability for this solution, then we look at the variational equation corresponding to (13.20). That is, let $w_1 = u_1 - u_2$ and $w_2 = v_1 - v_2$ be the difference of two solutions of (13.20). Then

$$w' = B(t)w \qquad (13.21)$$

where

$$B(t) = \begin{pmatrix} -\lambda_1 \dfrac{F(\lambda_1^{-1}u_1) - F(\lambda_1^{-1}u_2)}{u_1 - u_2} & \lambda \\ \lambda_2 b\,\dfrac{F(\lambda_1^{-1}u_1) - F(\lambda_1^{-1}u_2)}{u_1 - u_2} - \lambda_2 \dfrac{g(\lambda_1^{-1}u_1) - g(\lambda_1^{-1}u_2)}{u_1 - u_2} & -b \end{pmatrix}$$

There is a strong connection between A and B. In fact, writing $x = \lambda_1^{-1}u$, the conditions that $A(u)$ be column dominant with parameter δ are

$$\left.\begin{array}{l} -b + \lambda \le -\delta \\ \lambda^{-1} |b\, h(x) - b^2 - k(x)| + b - h(x) \le -\delta\,. \end{array}\right\} \quad (13.22)$$

Unraveling these inequalities, and setting $\lambda = b - \delta$ these become

$$\left.\begin{array}{l} 0 \le \delta < b \\ h(x) \ge b + \delta \\ \delta\, h(x) - \delta^2 \le k(x) \le (2b-\delta) h(x) + \delta^2 - 2b^2 \end{array}\right\} \quad (13.23)$$

On the other hand the values of k and h are contained in the values of g' and f respectively. Hence, if we apply the mean value theorem to each of thc inequalities they suggest

$$\left.\begin{array}{l} 0 \le \delta < b \\ f(x) \ge b + \delta \\ \delta\, f(x) - \delta^2 \le g'(x) \le (2b-\delta) f(x) + \delta^2 - 2b^2 \end{array}\right\} \quad (13.24)$$

for x and y in the given set. In fact, an integration of these from 0 to x yields (13.23). Now the remarkable thing is, that if we apply the same analysis to $B(t)$ we get the same conditions (13.24).

To be precise, we now assume that

$$\left.\begin{array}{l} 0 \le c_1 \le g'(x) \le c_2 \\ 0 < d_1 \le f(x) \le d_2. \end{array}\right\} \quad (13.25)$$

for all x. Then (13.24) is satisfied for $\delta = 0$ provided that $d_1 > b > 0$ and $c_2 \le 2b\, d_1 - 2b^2$. This can happen if and only if

$d_1^2 > 2c_2$. If this is satisfied, then there is a positive δ satisfying (13.24). Now we are ready to proceed. Let $|(u(0),v(0))|_1 \leq \|p\|\lambda_2 \delta^{-1}$. Then

$$\binom{u}{v}(t) = \Phi(t,0)\binom{u(0)}{v(0)} + \int_0^t \Phi(t,s)\binom{0}{\lambda_2 p(s)}\,ds \qquad (13.26)$$

where $\Phi(t,s)$ is the fundamental solution to

$$\binom{u}{v}' = A(u)\binom{u}{v}. \qquad (13.27)$$

Since A is column dominant with parameter δ we have $|\Phi(t,s)| \leq e^{-\delta(t-s)}$ if $t \geq s$. Applying this to (13.26) we have $|\binom{u}{v}(t)|_1 \leq \|p\|\lambda_2\delta^{-1}[e^{-\delta t} + \int_0^t \delta e^{-\delta(t-s)}ds] \leq \|p\|\lambda_2\delta^{-1}$. Thus the solution stays in the sphere $|u| + |v| \leq \|p\|\lambda_2\delta^{-1}$ and exists on R^+. By the usual arguments we get a solution on R. Since the variational equation is also column dominant, this bounded solution is exponentially stable so is almost periodic.

8. Notes. The maximum principle and its properties as well as the existence of bounded solutions in Example 13.7 may be found in Jackson, Subfunctions and second order ordinary differential inequalities Advances in Mathematics, 2(1968), 307-363.

The results of article 2 can be found in Fink, Almost automorphic and almost periodic solutions which minimize functionals, Tohoku Math. Journal 20(1968), 323-332.

The result 13.9 can be found in Seifert, Almost periodic solutions for systems of differential equations near points of nonlinear first approximations, Proc. Amer. Math. Soc. 11(1960), 429-435. The subsequent generalizations are in Fink, Uniqueness theorems and almost periodic solutions to second order equation, J. Differential Equations, 4(1968), 543-548, and Fink and Seifert, Liaponov functions and almost periodic solutions for almost periodic systems, J. Differential Equations, 2(1969), 307-313. Example 13.14 is in Langenhop and Seifert, Almost periodic solutions of second order nonlinear differential equations with almost periodic forcing, Proc. Amer. Math. Soc. 10(1959), 425-432.

The results on second order linear differential equations are a few of some very nice theorems in Markus and Moore, Oscillations and disconjugacy for linear differential equations with almost periodic coefficients, Acta. Math. 96(1956), 99-123. The use of these theorems for uniqueness theorems of nonlinear equations is in Fink, Uniqueness theorems and almost periodic solutions to second order equations, J. Differential Equations, 4(1968), 543-548 and Willett, Uniqueness for second order nonlinear boundary value problems with applications to almost periodic solutions, Ann. Mat. Pura Appl. 81(1969), 77-92.

The results of article 5 are in Frederickson and Lazer, Necessary and sufficient damping in a second order oscillator, J. Differential Equations, 5(1969), 262-270.

The application of column dominance to Lienard systems is in Fink, Convergence and almost periodicity of solutions of forced

Lienard equations, SIAM J. Appl. Math.

The Lienard equation has been widely studied, for boundedness theorems, see Graef, On the generalized Lienard equation with negative damping, J. Differential Equations, 12(1972), 34-62.

Chapter 14

Averaging

1. Introduction. Our previous applications of fixed point methods were in the case when the equation could be considered a perturbation of a linear equation for which we had an exponential dichotomy. If the equation has no obvious linear part then these methods do not apply. The question arises, is there a natural equation to which the given one can be related as a perturbation? A natural way to proceed is to find an autonomous equation with a solution around which an expansion will lead to a linear part that is dominant. The most obvious autonomous equation is the mean value equation.

2. The averaged equation. The idea that we propose is to be an extension of Theorem 8.4. So we look at the equation

$$x' = \epsilon\, f(x,t,\epsilon) \tag{14.1}$$

without the linear part that we had in that theorem. The averaged equation is

$$x' = \epsilon\, f_o(x) \tag{14.2}$$

where $f_o(x) = M_t\{f(x,t,0)\} = \lim_{T\to\infty}\frac{1}{T}\int_0^T f(x,t,0)\,dt$. In some sense we hope that solutions of (14.1) are close to those of (14.2). The latter equation should be in some sense be simpler since it is autonomous. And a natural type of solution to look for is a constant solution $x(t) \equiv x_o$ where $f(x_o) = 0$. In order to make the connection between (14.1) and such a solution one hopes to find an almost periodic change of variable $y = g(x,t,\epsilon)$ so that (14.1) becomes

$$y' = \epsilon\ [f_o(y) + f_1(y,t,\epsilon)] \tag{14.3}$$

with $f_1(y,t,0) = 0$. Then if $f_o(y_o) = 0$, we make the further change of variable $y = y_o + z$ so that the equation for z becomes

$$z' = \epsilon\ \left\{\left[\frac{\partial f_o}{\partial y}(y_o)\right]z + \left[f_o(y_o + z) - f_o(y_o) - \frac{\partial f_o}{\partial y}(y_o)\,z\right] + f_1(y_o + z,t,\epsilon)\right\} \tag{14.4}$$

and we can consider (14.4) as a perturbation of the linear system

$$z' = \epsilon\ \left[\frac{\partial f_o}{\partial y}(y_o)\right]z. \tag{14.5}$$

3. <u>Transformation Lemmas</u>. A glance at (14.1) and (14.3) shows that we want to integrate equations with right hand sides almost periodic with mean value zero. As we have seen before, this is not always possible. The idea of the compromise is to show that one can approximate in a reasonably nice way. We consider

$$x' = f(t) \tag{14.6}$$

where $M(f) = 0$. We do not know that this has a bounded integral. But we do know that for every $\eta > 0$

$$x' = -\eta x + f(t) \tag{14.7}$$

does. Then we try to take $\eta \to 0$ in (14.7) to study (14.6).

Lemma 14.1. Let F be a u.a.p. family with $M(f) = 0$ for all $f \in F$. There is a continuous function $a(\eta)$ on R^+ to R^+ with $a(0^+) = 0$ so that for every $f \in F$,

$$f_\eta(t) = \int_{-\infty}^{t} e^{-\eta(t-s)} f(s)\,ds \tag{14.8}$$

satisfies

(a) $|f_\eta(t)| \leq \frac{a(\eta)}{\eta}$

(b) $\|f'_\eta - f\| \leq a(\eta)$

(c) $f_\eta \in AP(E^n)$ and $\operatorname{mod}(f_\eta) \subset \operatorname{mod}(f)$

Proof: By definition f_η is a solution of (14.7) that is bounded so is in $AP(E^n)$ with module containment. Thus (b) and (c) follow from (a). Since the family F is u.a.p. and $M(f) = 0$ for all $f \in F$, $\lim_{T\to\infty} \frac{1}{T}\int_t^{t+T} f(s)\,ds = 0$ uniformly for $t \in R$ and $f \in F$. That is $\sup_{t,f}\left|\frac{1}{T}\int_t^{t+T} f(s)\,ds\right| = b(T)$ is bounded on R^+ and $\lim_{T\to\infty} b(T) = 0$. There exists a strictly decreasing differentiable function $\epsilon(T)$ on R^+ so that $\epsilon(0) > 1$, $\epsilon(T) \geq b(T)$ and $\lim_{T\to\infty} \epsilon(T) = 0$. Now if $f \in F$

$$|f_\eta(t)| = |\int_0^\infty e^{-\eta s} f(t-s)\,ds| = |\sum_{k=0}^\infty e^{-\eta kT} \int_{kT}^{(k+1)T} e^{-\eta(s-kT)} f(t-s)\,ds|$$

$$= |\sum_{k=0}^\infty e^{-\eta kT} \{ \int_{kT}^{(k+1)T} [e^{-\eta(s-kT)} - 1] f(t-s)\,ds - \int_{kT}^{(k+1)T} f(t-s)\,ds \}|$$

$$\leq \sum_{k=0}^\infty e^{-\eta kT} \int_{kT}^{(k+1)T} [1 - e^{-\eta(s-kT)}] \|f\| ds + \sum_{k=0}^\infty e^{-\eta kT} T\,\epsilon(T)$$

$$= T\,\epsilon(T)\,[1 - e^{-\eta T}]^{-1} + \|f\| \int_0^T [1 - e^{-\eta u}]\,du [1 - e^{-\eta T}]^{-1}$$

$$\leq T\,\epsilon(T)\,[1 - e^{-\eta T}]^{-1} + \|f\|\,T$$

since $1 - e^{-\eta u} \leq 1 - e^{-\eta T}$ on $[0,T]$. Now for each $\eta > 0$ $h(T) = \dfrac{\epsilon(T)}{1 - e^{-\eta T}}$ is a strictly decreasing function of T such that $h(0) > 1$ and $h(\infty) = 0$. Hence there is a unique $T(\eta)$ for which $h(T(\eta)) = 1$. Furthermore, T' exists and is negative. As $\eta \to 0$ $T(\eta)$ increases. If $T(0^+) \leq m$ then $\epsilon(T(\eta)) \geq \epsilon(m) > 0$ and hence $1 - e^{-\eta T(\eta)} = \epsilon(T(\eta)) > 0$ as $\eta \to 0$. This is not the case, so $T(0^+) = +\infty$ and thus $1 - e^{-\eta T(\eta)} \to 0$. It follows that $\eta T(\eta) \to 0$ as $\eta \to 0^+$. If we define $a(\eta) = \eta T(\eta)(1 + M)$ where $\|f\| \leq M$ for all $f \in F$, then $|f_\eta(t)| \leq \dfrac{a(\eta)}{\eta}$ and $a(0^+) = 0$. This completes the proof.

We want to apply this lemma to the case when the family F is $\{f(x,t) : x \in K\}$, K a compact set in E^n. For purposes of the

transformation of coordinates we need $f_\eta(x,t)$ to be smooth enough to allow differentiation. Note that by (14.8) the family f_η has the same continuity properties in x as f has. It is not in general a u.a.p. family since $\|f_\eta\|$ may not be bounded as $\eta \to 0$. But $\{\eta f_\eta\}_{\eta \geq 0}$ is a u.a.p. family as can be directly verified using (14.8). In fact
$|\eta f_\eta(t+\tau) - \eta f_\eta(t)| \leq \|(f(t+\tau) - f(t)\|$. This also shows that $\{f_\eta\}_{\eta \geq \eta_o > 0}$ is a u.a.p. family.

We now turn to the smoothness problem.

<u>Lemma 14.2</u>. Let $F = \{f(x,t) : x \in K\}$ be a u.a.p. family with $M(f) = 0$ for all $x \in K$. There is a continuous function $b(\eta)$ on R^+ to R^+ with $b(0^+) = 0$ and a function $u(x,t,\eta)$ with the following properties.

(a) $|u(x,t,\eta)| \leq b(\eta)\eta^{-1}$; $(x,t,\eta) \in K \times R \times (0,\infty)$

(b) $|\frac{\partial u}{\partial t}(x,t,\eta) - f(x,t)| < b(\eta)$; $(x,t,\eta) \in K \times R \times (0,\infty)$

(c) $u \in AP(E^n)$ uniformly for $(x,\eta) \in K \times [\eta_o,\infty)$, $\eta_o > 0$

(d) $\mathrm{mod}(u) \subset \mathrm{mod}(f)$.

(e) $\{\eta u(x,t,\eta\}_{\eta \geq 0}$ is a u.a.p. family. In fact
$\eta u(x,t,\eta) \to 0$ as $\eta \to 0$ uniformly on $K \times R$.

(f) If $\frac{\partial f}{\partial x}$ is uniformly continuous on $K \times R$, then
$\eta \frac{\partial u}{\partial x}(x,t,\eta) \to 0$ as $\eta \to 0$ uniformly on $K \times R$.

<u>Proof</u>: If $\eta > 0$ we can find trig polynomials $p_\eta(x,t)$ so that

$|f(x,t) - p_\eta(x,t)| < \eta$ on $K \times R$. In fact we may use Bochner-Fejer polynomials which also approximate $\frac{\partial f}{\partial x}$ if it exists and is almost periodic. We only use this latter observation for (f). We than use the fact that $G = F \cup \{p_\eta(x,t) : \eta > 0\}$ is a u.a.p. family and apply Lemma 14.1 to G, and note that $\operatorname{mod}(G) \subset \operatorname{mod}(F)$. Let $u(x,t,\eta) = \int_{-\infty}^{t} e^{-\eta(t-s)} p_\eta(x,s)\,ds$. Then we have (a) from Lemma 14.1. But

$$\left|\frac{\partial u}{\partial t}(x,t,\eta) - f(x,t)\right| \le \left|\frac{\partial u}{\partial t}(x,t,\eta) - p_\eta(x,t)\right| + \eta \le a(\eta) + \eta$$

thus for our purposes here we set $b(\eta) = a(\eta) + \eta$ and we have (a), (b), (c), and (d). Conclusion (e) follows as described above and estimate (b). Since we may take p_η so that $\left|\frac{\partial f}{\partial x}(x,t) - \frac{\partial p_\eta}{\partial x}(x,t)\right| < \eta$ if $\frac{\partial f}{\partial x}$ is almost periodic, conclusion (f) follows from the same sort of estimate as (a).

It is instructive to note that if $f(x,t)$ is a polynomial then $u(x,t,\eta)$ may be taken independent of η to be the integral of f, if f has the required derivatives.

Theorem 14.3. (Fundamental Lemma) Let $f(x,t,\epsilon)$ be almost periodic in t uniformly for $x \in K$ and ϵ fixed, and suppose f is continuous in three variables on sets of the form $K \times R \times [0,\epsilon_o]$ such that $f(x,t,\epsilon) \to f(x,t,0)$ as $\epsilon \to 0$ uniformly on $K \times R$. Further, suppose $\frac{\partial f}{\partial x}(x,t,\epsilon) \to \frac{\partial f}{\partial x}(y,t,0)$ as $(x,\epsilon) \to (y,0)$ uniformly in t, and define $f_o(x) = M_t\{f(x,t,0)\}$.

There exists a function $u(x,t,\epsilon)$ and $\epsilon_1 > 0$ such that

a) $u(x,t,\epsilon)$ is continuous on $K \times R \times [0,\epsilon_1]$, and almost periodic in t uniformly in $x \in K$ for fixed ϵ;

b) u, $\epsilon \frac{\partial u}{\partial x}$, $\frac{\partial u}{\partial t}(x,t,\epsilon) - f(x,t,0) + f_o(x)$ and $\frac{\partial}{\partial x}\left(\frac{\partial u}{\partial t}(x,t,\epsilon) - f(x,t,0) + f_o(x)\right)$ all $\to 0$ as $\epsilon \to 0$ uniformly on $K \times R$.

c) $\frac{\partial u}{\partial t}(x,t,\epsilon)$ is continuous on $K \times R \times [0,\epsilon_1]$ and has as many continuous derivatives with respect to x as one wants,

d) the transformation of variables

$$x = y + \epsilon\, u(y,t,\epsilon) \tag{14.9}$$

is invertible for $0 \leq \epsilon \leq \epsilon_1$ and transforms (14.1) into

$$y' = \epsilon\, f_o(y) + \epsilon\, F(y,t,\epsilon) \tag{14.10}$$

where $F(y,t,0) = 0$, $\frac{\partial F}{\partial y}(y,t,\epsilon) \to 0$ as $(y,\epsilon) \to (0,0)$ uniformly in t, and F satisfies the same properties as f. Furthermore $\text{mod}(F) \subset \text{mod}(f)$.

Proof: We write $g(x,t) = f(x,t,0) - f_o(x)$. Then $M_t(g(x,t)) = 0$ and we may apply Lemma 14.2. Let $u(x,t,\epsilon)$ be the function described in that Lemma. The properties (a), (b), and (c) follow from the estimates of Lemma 14.2. Now look at the transformation of variable (14.9). The Jacobian is $I + \epsilon\frac{\partial u}{\partial y}(y,t,\epsilon)$. Since $\epsilon\frac{\partial u}{\partial y}(y,t,\epsilon) \to 0$ as $\epsilon \to 0$ uniformly on $K \times R$, we take ϵ sufficiently small, say $< \epsilon_o$, so that this Jacobian is non-singular. Then the transformation is invertible with a differentiable inverse. At least this will be true locally. To be slightly more formal let $K_1 \equiv \{y \mid x = y + \epsilon\, u(y,t,\epsilon) \in K$ for $\epsilon \in [0,\epsilon_o]\}$ and for $\epsilon \in [0,\epsilon_o]$, $K_\epsilon \equiv \{x \mid x = y + \epsilon\, u(y,t,\epsilon), y \in K_1\}$. We note that there is a constant C so that

$|u(y,t,\epsilon) - u(z,t,\epsilon)| \leq C \sup\limits_{x\in K,\, t\in R} |\frac{\partial u}{\partial y}(x,t,\epsilon)|\, |y-z|$ by the mean value theorem. We pick ϵ_o so small that $C\,\epsilon\, |\frac{\partial u}{\partial y}| \leq \rho < 1$ on $K \times R \times [0,\epsilon_o]$. Then we claim that for $\epsilon \leq \epsilon_o$, K_ϵ is homeomorphic to K_1, and that the inverse function $y(x,t,\epsilon)$ is almost periodic in t uniformly on $K_1 \times R \times [\epsilon^*,\epsilon_o]$ for $\epsilon^* > 0$ and continuous on $K_1 \times R \times [0,\epsilon_o]$. To see this write

$x_1 = y(x_1,t,\epsilon) + \epsilon\, u(y(x_1,t,\epsilon),t,\epsilon)$ and

$x_2 = y(x_2,t,\epsilon) + \epsilon\, u(y(x_2,t,\epsilon),t,\epsilon)$ so that

$$y(x_2,t,\epsilon) - y(x_1,t,\epsilon) = x_1 - x_2 + \epsilon\, u(y(x_1,t,\epsilon),t,\epsilon) - \epsilon\, u(y(x_2,t,\epsilon),t,\epsilon)$$

and

$$|y(x_2,t,\epsilon) - y(x_1,t,\epsilon)| \leq |x_1 - x_2| + \epsilon\, C|u_y(\zeta,t,\epsilon)|\, |y(x_2,t,\epsilon) - y(x_1,t,\epsilon)|.$$

Thus

$$|y(x_2,t,\epsilon) - y(x_1,t,\epsilon)| \leq (1-\rho)^{-1}|x_1 - x_2|.$$

This shows that the mapping is 1 : 1 and continuous in x uniformly on $R \times [0,\epsilon_o]$. In the same way

$$|y(x,t,\epsilon) - y(x,s,\epsilon)| \leq \epsilon\ |u_t(x,\zeta,\epsilon)|\ |t-s|$$

so is continuous in t uniformly in x and ϵ. Finally

$|y(x,t+\tau,\epsilon) - y(x,t,\epsilon)| \leq$

$|\epsilon\ u(y(x,t+\tau,\epsilon),t+\tau,\epsilon) - \epsilon\ u(y(x,t,\epsilon),t+\tau,\epsilon)|$

$\quad + |\epsilon\ u(y(x,t,\epsilon),t+\tau,\epsilon) - \epsilon\ u(y(x,t,\epsilon),t,\epsilon)|$

$\quad \leq \rho|y(x,t+\tau,\epsilon) - y(x,t,\epsilon)| + v_{\epsilon u}(\tau)$. Thus $v_y(\tau) \leq (1-\rho)^{-1} v_{\epsilon u}(\tau)$. Now the family $\{\epsilon u\}$ is a u.a.p. family on sets of the form $K \times [0,\epsilon_1]$ so y has this property.

Now $(I + \epsilon\frac{\partial u}{\partial y})y' = x' - \epsilon\frac{\partial u}{\partial t}(y,t,\epsilon)$

$= \epsilon\ f(y+\epsilon u(y,t,\epsilon),t,\epsilon) - \epsilon\frac{\partial u}{\partial t}(y,t,\epsilon)$

$= \epsilon\ f_o(y) + \epsilon[f(y,t,0) - f_o(y) - \frac{\partial u}{\partial t}(y,t,\epsilon)]$

$\quad + \epsilon[f(y+\epsilon\ u(y,t,\epsilon),t,\epsilon) - f(y,t,0)] = \epsilon\ f_o(y) + \epsilon\ G(y,t,\epsilon)$.

The first bracket in G is (see first line of proof)

$g(y,t) - \frac{\partial u}{\partial t}(y,t,\epsilon)$ so $\to 0$ as $\epsilon \to 0$ uniformly in (y,t) so we define it to be 0 at $\epsilon = 0$. The second bracket is continuous at $\epsilon = 0$. Thus $G(y,t,0) = 0$. Clearly $G(y,t,\epsilon)$ is almost periodic uniformly on K_1 for $\epsilon < \epsilon_1$ and is continuous in three variables on $K_1 \times R \times [0,\epsilon_1]$, and $G(y,t,\epsilon) \to 0$ uniformly in t as $(y,\epsilon) \to (0,0)$. It is also easy to see that $\frac{\partial G}{\partial y}$ is continuous, and $\to 0$ uniformly in t as $(y,\epsilon) \to (0,0)$.

Now if ϵ_1 is sufficiently small, then $|\epsilon\frac{\partial u}{\partial y}(y,t,\epsilon) \leq \delta < 1$

(see (f) of Lemma 14.2). Thus

$$\left[I - \epsilon\frac{\partial u}{\partial y}(y,t,\epsilon)\right]^{-1} = \sum_{k=0}^{\infty} \left(-\epsilon\left(\frac{\partial u}{\partial y}(y,t,\epsilon)\right)\right)^k$$

is uniformly convergent so has the almost periodic and the continuity properties that G has. We now write the differential equation for y as

$$\begin{aligned} y' &= [I - \epsilon\frac{\partial u}{\partial y}(y,t,\epsilon)]^{-1}[\epsilon\, f_o(y) + \epsilon\, G(y,t,\epsilon)] \\ &= \left(I + \sum_{k=1}^{\infty} \epsilon^k\left(\frac{\partial u}{\partial y}(y,t,\epsilon)\right)^k\right)(\epsilon\, f_o(y) + \epsilon\, G(y,t,\epsilon)) \\ &= \epsilon\, f_o(y) + \epsilon\, F(y,t,\epsilon) \end{aligned}$$

and we see that F has the required properties.

Note that we have not only reduced the equation (14.1) to (14.3) so that the transformation from (14.3) to (14.1) is almost periodic. But the reverse transformation is almost periodic so that in some sense we have equivalent problems.

4. The Main Theorem. We are now in a position to gather all the facts together

Theorem 14.4. Let $f(x,t,c)$ be almost periodic in t uniformly on any compact set $K \times \{\epsilon\}$. Suppose f is continuous as a function of three variables such that $f(y,t,\epsilon) \to f(x,t,0)$

uniformly in t as $(y,\epsilon) \to (x,0)$. Suppose $\frac{\partial f}{\partial x}(x,t,\epsilon) \to \frac{\partial f}{\partial x}(y,t,0)$ as $(x,\epsilon) \to (y,0)$ uniformly in t. If $f_o(x) = M_t\{f(x,t,0)\}$ is such that there is an $x_o \in \text{int}(K)$ with $f_o(x_o) = 0$ and $\frac{\partial f_o}{\partial x}(x_o)$ having no eigenvalues with zero real part, then there is an $\epsilon_o > 0$ so that for $0 \leq \epsilon \leq \epsilon_o$ there is an almost periodic solution $x(t,\epsilon)$ to (14.1) with

a) $x(t,0) = x_o$;

b) $x(t,\epsilon)$ is almost periodic with $\text{mod}(x) \subset \text{mod}(f)$;

c) If $f(x,t,\epsilon)$ is u.a.p. on $K \times [\epsilon^*,\epsilon_o]$ if $\epsilon^* > 0$, then $x(t,\epsilon)$ is continuous as a function of two variables on $R \times [0,\epsilon_o]$; and

d) If all the eigenvalues of $\frac{\partial f_o}{\partial x}(x_o)$ have negative real part, then $x(t,\epsilon)$ is uniformly asymptotically stable.

Proof: We use Theorem 14.3 to reduce (14.1) to (14.10) as the first step, where F satisfies the same hypothesis as f except that $F(y,t,0) \equiv 0$, and $\frac{\partial F}{\partial y}(y,t,\epsilon) \to 0$ as $(y,\epsilon) \to (0,0)$ uniformly in t. Then make the change of variable $y = z + y_o$ to get

$$z' = \epsilon Az + \epsilon h(z,t,\epsilon) \tag{14.11}$$

where $A = \frac{\partial f_o}{\partial x}(x_o)$ and

$$h(z,t,\epsilon) = f_o(y_o + z) - f_o(y_o) - \frac{\partial f_o}{\partial y}(y_o)z + F(z + y_o, t, \epsilon).$$

Note that $h(z,t,\epsilon)$ satisfies the same hypothesis as f and $\mathrm{mod}(h) \subset \mathrm{mod}(f)$, with $h(z,t,\epsilon)$ and $\frac{\partial}{\partial z}h(z,t,\epsilon) \to 0$ uniformly in t as $(z,\epsilon) \to (0,0)$. As a final change of variables, let $w(t) = z(\epsilon^{-1}t)$. Then

$$w' = Aw + h(w, \epsilon^{-1}t, \epsilon). \qquad (14.12)$$

To prove the existence of an almost periodic solution of (14.12), let $r > 0$ be such that $|z| \leq r$ implies that $y = z + y_o \in K$. Then $|w| \leq r$ also implies this. Since $x' = Ax$ satisfies an exponential dichotomy let the norm of the mapping from f to the bounded solution of $x' = Ax + f(t)$ be M. We pick $\epsilon_o < \epsilon_1$ where ϵ_1 is small so Theorem 14.3 applies and ϵ_o is sufficiently small so that $\sup|h(w,\epsilon^{-1}t,\epsilon)|M \leq r$ where the sup is taken over $|w| \leq r$, $t \in R$ and $0 \leq \epsilon \leq \epsilon_o$. This is the possible since $h(w,t,\epsilon) \to 0$ as $\epsilon \to 0$ uniformly in the other variables. Thus if φ is almost periodic sith $\|\varphi\| \leq r$ and $\mathrm{mod}(\varphi) \subset \mathrm{mod}(h)$ we have a unique solution $T\varphi$ of $w' = Aw + h(\varphi(t), \epsilon^{-1}t, \epsilon)$ which is almost periodic such that $\mathrm{mod}(T\varphi) \subset \mathrm{mod}(h)$ and $\|T\varphi\| \leq M\,\|h(\varphi(t),\epsilon^{-1}t,\epsilon)\| \leq r$. This all holds for any $0 \leq \epsilon \leq \epsilon_o$. Now in order to make T a contraction we note that $\frac{\partial h}{\partial w} \to 0$ as ϵ and $r \to 0$ uniformly in the other variable so take ϵ_o and r sufficiently small to get

$\delta = M \sup|\frac{\partial h}{\partial w}(w,\epsilon^{-1}t,\epsilon)| < 1$ on $|w| \leq r$, $0 \leq \epsilon \leq \epsilon_o$. In actuality we must do this first, then fixing r take ϵ_o smaller so the first requirement is satisfied. In any case, we now have a unique fixed point of T in $|w| \leq r$ as long as $0 \leq \epsilon \leq \epsilon_o$, say $\varphi(t,\epsilon)$ is the fixed point. Note that $\|\varphi(t+\tau,\epsilon) - \varphi(t,\epsilon)\|$ $\leq M\|h(\varphi(t+\tau,\epsilon),\epsilon^{-1}(t+\tau),\epsilon) - h(\varphi(t,\epsilon),\epsilon^{-1}t,\epsilon)\|$ $\leq M(\sup \frac{\partial h}{\partial w})\|\varphi(t+\tau) - \varphi(t)\| + M\|h(\varphi(t,\epsilon),\epsilon^{-1}(t+\tau),\epsilon)$ $- h(\varphi(t,\epsilon),\epsilon^{-1}t,\epsilon)\|$. Therefore

$$\|\varphi(t+\tau,\epsilon) - \varphi(t,\epsilon)\| \leq (1-\delta)^{-1}M\|h(\varphi(t,\epsilon),\epsilon^{-1}(t+\tau),\epsilon) - h(\varphi(t,\epsilon),\epsilon^{-1}t,\epsilon)\|.$$

Letting $\psi(t,\epsilon) = \varphi(\epsilon t,\epsilon)$, we have that ψ is an almost periodic solution of (14.11) such that $\|\psi(t+\tau,\epsilon) - \psi(t,\epsilon)\| \leq (1-\delta)^{-1}M\, v_h(\tau)$, $0 \leq \epsilon \leq \epsilon_o$ where v_h is the Bochner translation function for h on $|w| \leq r$, $[0,\epsilon_o]$. The function v_h is almost periodic if f is u.a.p. on $K \times [\epsilon^*,\epsilon_o]$, $\epsilon^* > 0$. This follows since h is u.a.p. on $K \times [c^*,\epsilon_o]$ by that property of f. But since $h \to 0$ uniformly as $\epsilon \to 0$, this extends to $[0,\epsilon_o]$. Thus the family $\{\psi(t,\epsilon),\ 0 \leq \epsilon \leq \epsilon_o\}$ is a u.a.p. family and as $\epsilon_n \to \epsilon^*$, we can extract a subsequence ϵ_n' so that $\psi(t,\epsilon_n') \to \psi_o(t)$ a solution of the ϵ^* equation in $|w| \leq r$. But the contraction gives a unique solution there, so $\psi_o(t) = \psi(t,\epsilon^*)$. Thus continuity follows and (c) is verified. Clearly $\psi(t,0) = 0$ so (a) is satisfied. Also (d) follows easily.

5. <u>van der Pol Example</u>. The standard example for this sort of thing is the van der Pol equations. The equation is

$$y'' - \epsilon(1 - y^2)y' + y = A \sin \omega_1 t + B \sin \omega_2 t \tag{14.12}$$

where ϵ, ω_1, and ω_2 are positive parameters, A and B are real numbers and we are looking for almost periodic solutions. What one does is first write the system as $x_1 = y$ and $x_2 = y'$ to get

$$\left.\begin{aligned} x_1' &= x_2 \\ x_2' &= \epsilon(1 - x_1^2)x_2 - x_1 + A \sin \omega_1 t + B \sin \omega_2 t. \end{aligned}\right\} \tag{14.13}$$

One attempts to use variation of parameters to solve the $\epsilon \neq 0$ equation from the $\epsilon = 0$ solution. Thus we take

$$\begin{aligned} x_1 &= u_1 \cos t + u_2 \sin t + A(1 - \omega_1^2)^{-1} \sin \omega_1 t + B(1 - \omega_2^2)^{-1} \cos \omega_2 t \\ x_2 &= - u_1 \sin t + u_2 \cos t + A\omega(1 - \omega_1^2)^{-1} \cos \omega_1 t + B\omega_2 (1 - \omega_2^2)^{-1} \sin \omega_2 t \end{aligned} \tag{14.14}$$

and note that for $\epsilon = 0$, this is a solution with constant u_1 and u_2. For $\epsilon \neq 0$ put (14.14) into (14.13) and consider u_1 and u_2 as functions. Then (x_1, x_2) is a solution of (14.13) if and only if (u_1, u_2) is a solution of

$$\left.\begin{aligned} u_1' &= \epsilon(1 - x_1^2)x_2 \sin t \\ u_2' &= \epsilon(1 - x_1^2)x_2 \cos t \end{aligned}\right\} \tag{14.15}$$

with x_1 and x_2 as in (14.14). We would like to apply Theorem

14.4. In this case $f(u,t,\epsilon) = (1-x_1^2)x_2\binom{\sin t}{\cos t}$ and is independent of ϵ, and has continuous first partials with respect to u_1 and u_2, in fact continuous uniformly in t. Thus the continuity hypotheses of Theorem 14.4 are satisfied. It remains to find f_o and a zero of f_o. This function f_o depends on ω_1 and ω_2. If $m + m_1\omega_1 + m_2\omega_2 \neq 0$ for integers m, m_1, and m_2 such that $|m| + |m_1| + |m_2| \leq 4$, then

$$f_o(u_1,u_2) = \frac{1}{8}[2(2 - A^2(1-\omega_1^2)^{-2} - B^2(1-\omega_2^2)-2) - u_1^2 - u_2^2]\binom{u_1}{u_2}$$

the equation $u' = f_o(u_1,u_2)$ has the solution $u_1 = u_2 = 0$ and $\left(\frac{\partial f_o}{\partial u}\right)_{u_i=o} = \text{diagonal } (\frac{1}{4}[2 - A^2(1-\omega_1^2)^{-2} - B^2(1-\omega_1^2)^{-2}])$. Thus if this diagonal matrix is not zero we can apply the theorem to get an almost periodic solution to (14.12) which is continuous in ϵ and is near $y = A(1-\omega_1^2)^{-1}\sin\omega_1 t + B(1-\omega_2^2)^{-1}\sin\omega_2 t$. If the diagonal matrix is negative, then this solution is uniformly asymptotically stable.

6. Duffing Example. The second example is the Duffing equation

$$y'' + (1-\epsilon\upsilon)y + \epsilon^3 y^3 = p(t). \tag{14.16}$$

In order to apply averaging, one writes $x = \epsilon y$ to get

$$x'' + x - \epsilon\upsilon x + \epsilon x^3 = \epsilon p(t) \tag{14.17}$$

which we write as a system with $z_1 = x$ and $z_2 = x'$ so that $z = (z_1, z_2)^T$ satisfies

$$z' = (A + \epsilon C)z + \epsilon g(z) + \epsilon f(t) \tag{14.18}$$

where

$$A = \begin{pmatrix} 0 & 1 \\ -1 & 0 \end{pmatrix}, \qquad C = \begin{pmatrix} 0 & 0 \\ \upsilon & 0 \end{pmatrix}, \qquad g(z) = -\begin{pmatrix} 0 \\ z_1^3 \end{pmatrix},$$

and $f(t) = \begin{pmatrix} 0 \\ p(t) \end{pmatrix}$. A further change of variable $\omega = e^{-At}z$ leads to

$$\omega' = \epsilon e^{-At} C e^{At}\omega + \epsilon e^{-At} g(e^{At}\omega) + \epsilon e^{-At} f(t). \tag{14.19}$$

Note that e^{-At} is almost periodic, so that (14.19) is an almost periodic equation of the form $\omega' = \epsilon f(\omega, t, \epsilon)$ with $f(\omega, t, \epsilon)$ independent of ϵ and satisfying the continuity requirements of Theorem 14.4. The function $f_o(\omega)$ is easily computed. It is

$$f_o(\omega) = \begin{pmatrix} 0 & -\upsilon/2 \\ \upsilon/2 & 0 \end{pmatrix}\begin{pmatrix} \omega_1 \\ \omega_2 \end{pmatrix} - \frac{3}{8}\left(\omega_1^2 + \omega_2^2\right)\begin{pmatrix} \omega_2 \\ \omega_1 \end{pmatrix} + \frac{1}{2}\begin{pmatrix} -b_1 \\ a_1 \end{pmatrix}$$

where a_1 and b_1 are the fourier coefficients of p when p is written in its real form

$$p(t) \sim \sum_{n=1}^{\infty} (a_n \cos \lambda_n t + b_n \sin \lambda_n t)$$

with $\lambda_1 = 1$. Note that by a translation of axis $t \to t + \alpha$, the only term in (14.16) that changes is p, hence we may assume that $b_1 = 0$, but we will require that $1 \in \exp(p)$ so that $a_1 \neq 0$. Assume $a_1 > 0$. With this in mind we look for zeros of $f_o(\omega)$ when $\omega_2 = 0$. Thus we need find a zero of the equation

$$\upsilon/2\ \omega_1 - 3/8\ \omega_1^3 + 1/2a = 0 \qquad (14.20)$$

If $\upsilon > \left(\frac{9a_i}{2}\right)^{2/3}$ then (14.20) has three real solutions ω_1, one of which is negative and $4\upsilon/9 < \omega_1^2 < 4\upsilon/3$. Let this value be $\bar{\omega}_1$. Then $\frac{\partial f_o}{\partial \omega}(\bar{\omega}_1, 0)$ is $\begin{pmatrix} 0 & -\upsilon/2 + 3\bar{\omega}_1^2/8 \\ \upsilon/2 - 9\bar{\omega}_1^2/8 & 0 \end{pmatrix}$ so has non-zero real eigenvalues. Thus we can conclude that for ϵ small, (14.16) has almost periodic solutions which are in fact large for small ϵ. This is because (14.16) has resonance when $\epsilon = 0$. Of course the almost periodic solution to (14.17) that one gets is near the $\epsilon = 0$ solution which is

$\begin{pmatrix} x_1 \\ x_2 \end{pmatrix} = e^{At}\bar{\omega}$, with $\bar{\omega} = \begin{pmatrix} 0 \\ \bar{\omega}_1 \end{pmatrix}$.

The above idea can be extended to a general result about equations of the form $x' = (A + \epsilon\, C(t))x + \epsilon\, g(x,\epsilon) + \epsilon\, p(t)$, with g being a polynomial of odd degree and C and A of appropriate form.

7. Two timing example. It is not necessary to average (14.1) completely. If f can be written in two pieces, then it is

possible to consider one piece as a perturbation of the other. Specifically, consider an equation of the form

$$x' = \epsilon f(x,t) + \epsilon g(x,\epsilon t). \tag{14.21}$$

Here the fact that the argument of g is ϵt denotes the slow time as ϵ is small. The idea will be to average only over the fast time f. Thus we let $f_o(x) = M_t\{f(x,t)\}$ when f is almost periodic uniformly for x in compact sets, and use Lemma 14.3 with the function $f(x,t) - f_o(x)$. Call that function $u(x,t,\epsilon)$ as before. Then (14.21) is transformed by $x = y + \epsilon u(y,t,\epsilon)$ into

$$y' = \epsilon\left(I + \epsilon\frac{\partial u}{\partial y}(y,t,\epsilon)\right)^{-1}\Big\{f_o(y) + \left[-\frac{\partial u}{\partial t}(y,t,\epsilon) + f(y,t) - f_o(y)\right]$$

$$+ \left[f(y + \epsilon u(y,t,\epsilon),t) - f(y,t)\right] + g(y,\epsilon t)$$

$$+ [g(y + \epsilon u(y,t,\epsilon),\epsilon t) - g(y,\epsilon t)]\Big\}.$$

Write this as

$$y' = \epsilon f_o(y) + \epsilon g(y,\epsilon t) + \epsilon h(y,t,\epsilon). \tag{14.22}$$

Note that $h(y,t,\epsilon)$ satisfies the conditions that

(a) $h(y,t,\epsilon) \to 0$ as $(y,\epsilon) \to 0$ uniformly in t;

(b) $\frac{\partial h}{\partial y}(y,t,\epsilon) \to 0$ as $(y,\epsilon) \to 0$ uniformly in t; and

(c) h is almost periodic in t uniformly for y in compact sets;

provided H_1 : f and g are almost periodic in t uniformly for y in compact sets; and $f(y,t)$, $\frac{\partial f}{\partial y}(y,t)$, $g(y,t)$, and $\frac{\partial g}{\partial y}(y,t)$, are continuous in y uniformly in t.

We write $G(y,\epsilon t) = f_o(y) + g(y,\epsilon t)$. We will assume that $g(y,t)$ is periodic in t with period L; and consider the equation

$$y' = \epsilon G(y,\epsilon t). \tag{14.23}$$

In both (14.22) and (14.23) let $z(t) = y(\epsilon^{-1}t)$ to get

$$z' = G(z,t) + h(z,\epsilon^{-1}t,\epsilon) \qquad (14.22)'$$

and

$$z' = G(z,t). \qquad (14.23)'$$

We want to consider (14.22)' as a perturbation of (14.23)' and to this end we assume

H_2: There is a periodic solution y_o of (14.23)' with period L such that the variational equation

$$\omega' = \frac{\partial G}{\partial \omega}(y_o(t),t)\omega$$

has characteristic exponents with non-zero real parts. (Or equivalently satisfies an exponential dichotomy.)

We then make the change of variables $\omega = z - y_o(t)$ so that (14.22)' becomes

$$\omega' = A(t)\omega + [G(\omega + y_o, t) - G(y_o, t) - A(t)\omega] + h(\omega + y_o, \epsilon^{-1}t, \epsilon) \tag{14.24}$$

which can be written as

$$\omega' = C\omega + H(\omega, t, \epsilon) \tag{14.25}$$

by using Floquet Theory where the fundamental solution of $\omega' = A(t)\omega$ is $P(t)e^{Ct}$ and $H(\omega,t,\epsilon) = P(t)[G(p(t)\omega + y_o, t) - G(y_o, t) - A(t)P(t)\omega] + h(P(t)\omega + y_o, \epsilon^{-1}t, \epsilon)$. Now (14.25) can be solved as in Theorem 14.4 since the contraction mapping can be used for (ω, ϵ) small. After tracing this all back one discovers we get an almost periodic solution $x(t,\epsilon)$ of (14.21) such that $\lim_{\epsilon \to 0^+} |x(t,\epsilon) - y_o(\epsilon t)| = 0$ uniformly in t.

Summarizing, if H_1 and H_2 are satisfied, then (14.21) has an almost periodic solution $x(t,\epsilon)$ such that $\lim_{\epsilon \to 0^+} |x(t,\epsilon) - y_o(\epsilon t)| = 0$ uniformly in t.

8. Notes. The method of averaging is due to Krylov and Bogoliuboff. A thorough discussion of the method can be found in Bogoliuboff and Mitropolski, Asymptotic methods in the theory of nonlinear oscillations, Gordon and Breach, 1962. There are two different problems, the one on constructing solutions on finite times, and the one we have discussed here on R. The latter is more difficult. The form of Theorem 14.4 is in Hale, Ordinary Differential equations, Wiley, 1969. The application to the

van der Pol equation is also in that book.

The Duffing equation example is in Seifert, On almost periodic solutions for undamped systems with almost periodic forcing, Proc. Amer. Math. Soc. 31(1972), 104-108. It is generalized there and in Seifert, Almost periodic solutions by the method of averaging, Japan-United States Seminar on Ordinary Differential and Functional Equations. Lecture Notes 243, Springer-Verlag (1973).

The fast time slow time application is in Sethna, An extension of the method of averaging, Quart. Appl. Math. 25(1967), 205-211.

Chapter 15

THE LITERATURE

1. Citations. We give here the references specifically cited in the text and not in the bibliography. A series of numbers appear after each reference. The first gives a reference to a review. The key is:

A. MR x, y means Mathematical Reviews, volume x, page y or review y.

B. Z, x, y means Zentralblatt. volume x. page y or review y.

C. F d. m., x, y means Fortscritte der Mathematik, volume x, page y.

The second set of symbols are the section numbers where these references are mentioned in the text.

[1] Bochner, S., Beiträge zur Theorie der fastperiodische Funktionen. I : Funktionen einer Variablen, Math. Ann. 96(1927), 119-147. F d. m. 52, 261. 1.14.2.8.

[2] Jessen, B., On the proofs of the Fundamental Theorem on Almost Periodic Functions, Det. KGL Danske Viden. Selskob Mat-Fys. Meddel. XXV(1949), 1-12. MR 11,101. 3.9.

[3] Bochner, S., Properties of Fourier Series of almost periodic functions, Proc. London Math. Soc., 26(1927), 433-452. F d. m. 53, 254. 4.9.

[4] Jessen, B., Some aspects of the theory of Almost Periodic Functions, Proc. Int. Congress Amsterdam 2(1954), 1-10. 6.10.

[5] Coppel, W. A., Stability and Asymptotic behavior of Differential Equations, Heath Mathematical Monographs, 1965. 7.8.

[6] Markus, L., and Yamake, Global stability criteria for differential systems, Osaka Math. J. 12(1960), 305-317. 7.8.

[7] Sandberg, I., Some theorems on the Dynamic response of nonlinear transistor networks, B.S.T.J. 48(1969), 35-54. 7.8.

[8] Fréchet, M., Les fonctions asymptotiquement presque-périodiques, Revue Sci. (Rev. Rose Illus.). 79,341-354. (1941). MR 7, 127; Les fonctions asymptotiquement presque-périodiques continues, C. R. Acad. Sci. Paris 213, 520-522. (1941). MR 5, 96. 9.7.

[9] Meisters, G. H., On almost periodic solutions of a class of differential equations, Proc. Amer. Math. Soc. 10(1959), 113-119. 9.7.

[10] Meisters, G. H., and Fink, A. M., On Δ-m sets and group topologies, Rocky Mount. J. 2(1972), 225-233. 11.11.

[11] Sell, G., Periodic solutions and asymptotic stability, J. Differential Equations 2(1966), 143-157. 11.11.

[12] Denjoý, A., Sur les courbes defines par les équations differéntielles a le surface du tore, J. de Math. 11(1932), 333-375. 12.7.

[13] Coddington, E., Levinson, N., Theory of Ordinary Differential Equations, McGraw Hill, 1955. 12.3.

[14] Bhatia, N. P., and Szegö, G. P., Dynamical Systems: Stability Theory and Applications, Springer, Berlin, 1967. 12.4.

[15] Jackson, L., Subfunctions and second order ordinary differential inequalities Advances in Mathematics, 2(1968), 307-363. 13.2, 13.8.

[16] Schmitt, K., Periodic solutions of nonlinear second order differential equations, Math. Z., 98(1967), 200-207. 13.2.

[17] Heimes, K. A., Boundary value problems for ordinary nonlinear second order systems, J. Differential Equations, 2(1966), 449-463. 13.2.

[18] Reuter, G. E. H., On certain non-linear differential equations with almost periodic solutions. London. Math. Soc. Journal 26(1951), 215-221. 13.6.

[19] Fink, A. M., Convergence and almost periodicity of solutions of forced Lienard equations, SIAM J. Appl. Math. 13.8.

[20] Graef, J. R., On the generalized Lienard equation with negative damping, J. Differential Euqations, 12(1972), 34-62. 13.8

[21] Hale, J., Ordinary Differential equations, Wiley, 1969. 13.8.

2. Bibliography. The papers given here are related to almost periodic differential equations. The numbers after each reference have the meanings as in the previous section. A major portion of this bibliography was gathered by my colleagues G. H. Meisters and P. O. Fredcrickson, the former for the years 0-1958 and the latter for the years 1958-1964.

1. Abel, J. On the almost periodic Mathieu equation. Quart. of Appl. Math. 28(1970) 205-217. MR 43, 623.

2. _______, Uniform almost orthogonality and the instabilities of an almost periodic parametric oscillator. J. Math. Anal. Appl. 36(1971) 110-122. MR 43, 6520.

3. Adrianova, L. J. The reducibility of systems of n linear differential equations with quasi-periodic coefficients. (Russian). Vestnik Leningrad. Univ. 17(1962), no. 7, 14-24. MR 25, 3208.

4. Alimzanova, R. M., and Zolotarev, J. G. The transformation of analytic systems of differential equations with quasi-periodic coefficients. (Russian) Kazah. Gos Ped. Inst. Ucen. Zap. 23(1966) 45-54. MR 37, 3111.

5. Alymkulov, K. The construction of almost periodic solutions of certain differential equations. (Russian). Studies in Integro-Differential Equations in Kirghizia, No. 6. (Russian). pp. 226-231. Izdat. "Ilim", Frunze, 1969. MR 45, 7236.

6. Amerio, L. Soluzioni quasi-periodiche, o limitate, di sistemi differenziali non lineari quasi-periodici, o limitati. Annali mat. pura ed appl. 39(1955) 97-119, 6.10, 10.5.

7. _______, Funzioni quasi-periodiche ed equazioni differenziali. Atti VI Congr. Un. Mat. Ital. (Naples, 1959), pp. 119-144. Edizioni Cremonese, Rome, 1960. MR 23, 4389.

8. _______, Sull'integrazione delle funzioni quasi-periodiche a valori in uno spazio hilbertiano. Rend. Accad. Naz. Lincei (8) 28(1960) 600-603. MR 23, A 2003.

9. _______, Bounded or almost-periodic solutions of non-linear differential systems. Proceedings of the Conference on Differential Equations. pp. 179-82. Univ. of Maryland Book Store. MR 18, 738.

_______, Quasi-periodicità degli integrali ad energia limitata dell'equazione delle onde, con termine noto quasi-periodico. I, II, III. Rend. Accad. Naz. Lincei (8) 28(1960) 147-152, 322-327, 461-466. MR 22, 6927.

_______, Sull'equazione delle onde con termine noto quasi-periodico. Rend. Mat. Appl. (5) 19(1960) 333-346. MR 24, A 1505.

_______, Problema misto e quasi-periodicita per l'equazione delle onde non omogenes. Ann. Mat. pura ed appl. (4) 49(1960) 393-417. MR 22, 8210.

_______, Sulle equazioni differenziali quasi-periodiche astratte. Ric. Mat. 9(1960) 256-274. MR 24, A 896.

_______, Problema misto e soluzioni quasi-periodiche del-l'equazione delle onde. (English summary) Rend. Sem. Mat. Fis. Milano 30(1960) 197-222. MR A 1506.

_______, Sull'integrazione delle funzioni quasi-periodiche astratte. Ann. di Mat. 53(1961) 371-382. MR 24, A 807.

_______, Ancora sulle equazioni differenziali quasi-periodiche. Ric. Mat. 10(1961) 31-32. MR 24, A 3386.

_______, Funzioni quasi-periodiche astratte e problemi di propagazione. Conferenze del Seminario di Matematica dell'Universita di Bari, no. 57(1961) 1-14. MR 32, 6058.

_______, Sulle equazioni lineari quasi-periodiche negli spazi hilbertiani. I, II. Rend. Accad. Naz. Lincei 31(1961) 110-117, 197-205. MR 25, 3389, MR 26, 2879.

_______, Soluzioni quasi-periodiche delle equazioni lineari iperboliche quasi-periodiche. Rend. Accad. Naz. Lincei (8) 33(1962) 179-186. MR 28, 1399.

_______, Soluzioni quasi-periodiche di equazioni funzionali negli spazi di Hilbert. Seminario dell'Istituto Nazionale di Alta Matematica (1962-1963) 787-796. MR 33, 445.

_______, Su un teorema di minimax per le equazioni differenziali astratte. Rend. Accad. Naz. Lincei 35(1963) 409-416. MR 29, 5126.

_______, Soluzioni quasi-periodiche di equazioni quasi-periodiche negli spazi hilbertiani. Ann. Mat. Pura Appl. (4) 61(1963) 259-277. MR 27, 6160.

Sull'integrazione delle funzioni $l^p\{X_n\}$ quasi-periodiche, con $1 \leq p < +\infty$. Ric. di Mat. 12(1963) 3-12. MR 27, 6159.

24. _______, Solutions presque-periodiques d'equations fonctionnelles dans les espaces de Hilbert. Deuxième Colloque sur l'Analyse fonctionnelle, Liège. 1964 11-35. M32, 1575.

25. _______, Solujioni quasi-periodiche di equatzioni funzionali lineari e non lineari Comptes Rendus de la IIIe Reunion du Groupement des Math. d'Expressione Latine pp. 15-33(1965) Naumur. MR 39, 1783.

26. _______, Almost-periodic solutions of the equation of Schrödinger type. Atti Accad. Nay Lincei Rend. Cl. Sci. Fiz. Mat. Natur 8(43) (1967) 147-153. MR 39, 1284, 265-270, MR 38, 3566.

27. Amerio, L., and Prouse, G. Bounded or almost-periodic solutions of a nonlinear wave equation. I, II. Rend. Acc. Naz. Lincei (1966) 40-41.

28. _______, On the non-linear wave equation with dissipative term discontinuous with respect to the velocity, I, II. (Italian summary) Atti Accad. Naz. Lincei Rend. Cl. Sci. Fis. Mat. Natur. (8) 44(1968) 491-496; ibid. (8) 44(1968) 615-624. MR 39, 1813.

29. _______, Uniqueness and almost-periodicity theorems for a non-linear wave equation. (Italian summary). Atti Accad. Naz. Lincei Rend. Cl. Sci. Fis. Mat. Natur. (8) 46(1969) 1-8. MR 41, 653.

30. _______, Almost-periodic functions and functional equations. (The University Series in Higher Mathematics.) New York etc.: Van Nostrand Reinhold Company 1971, VIII, 184 p. Ł 7.00. Z 215, 157.

31. Aramescu, C. Sur l'existence des solutions presque-periodiques des systems differentielles du premier ordre. Atti Accad. Naz. Lincei, Cl. Sci. Fiz. Mat. Natur. (8) 47(1969) 468-471. MR 43, 624.

32. Ararkyian, B. G. Departure to almost periodic behavior of solutions of boundary value problems for a hyperbolic equation. Steklov Inst. Proc. 103(1968) 1-12. MR 43, 722.

33. _______, On the asymptotic almost-periodicity of solutions of some nonstationary equations. Soviet Math. 13(1972) 943-945.

34. Arnol'd, V. I. Generation of quasi-periodic motion from a family of periodic motions. (Russian). Dokl. Akad. Nauk SSSR 138(1961) 13-15. MR 24, A801. Translated in Soviet Math. Dokl. 2 247-249.

35. _______, Proof of a Theorem of A. N. Kolmogorov on the preservation of conditionally periodic motions under a small perturbation of the Hamiltonian. (Russian). Uspehi Mat. Nauk 18(1963) no. 5(113) 13-40. MR 29, 328, MR 30, 943.

36. _______, Small denominators, I, on the mapping of a circle into itself. Transl AMS Series 2 vol. 46, 213-284.

37. Arnol'd, V. I., and Avez. Ergodic problems of classical mechanics. Benjamin, N.Y. (1968).

38. Artjušenko, L. M. The application of Fourier series for finding almost periodic solutions of equations with mean values. (Russian). Izv. Vyss. Ucebin. Zaved. Matematika, no. 4 (5)(1968) 21-27. MR 25, 2382.

39. _______, Zur frage der fastperiodischen Lösungen von Gleichungen mit abweichenden Argument. Diff. Urav. 6(1970) 1389-1394. Z 196, 112.

40. Avakian, A. Almost periodic functions and the vibrating membrane. J. Math. Phys. XIV(1935) 350-378. Z 16, 263.

41. Bailey, P., and Shampine, L. Concerning periodic solutions of $y'' = f(t,y,y') = 0$. J. Math. Anal. Appl. 23(1968) 558-574. MR 37, 6542.

42. Barbalat, I. Application du principe topologique de T. Ważewski aux équations différentielles du second ordre. Annales Polonici Mathematici, V(1958) 303-317. MR 21, 4277.

43. _______, Solutions presque-périodiques des equations différentielles non-linéaires. Com. Acad. R. S. Romainia, 11(1961) 155-159. MR 24, 292.

44. _______, Solutions presque-périodiques de l'équation de Riccati. Com. Acad. R. S. Romainia. 11(1961) 161-165. MR 24, 294.

45. _______, Systems d'equations differentielles de type Lienard. Abh. Deutsch. Ahad. Wiss. Berlin Kl. Math. Phys. Tech. (1965) 206-218. MR 33, 4382.

46. Barbalat, I., and Halanay, A. Applications of the frequency method to forced nonlinear oscillations. Math. Nacha. 44(1970) 165-179.

47. Barbu, V. Solutions presque-périodiques pour un systeme d'equations lineaires aux derivées partielles. Ric. Mat. XV(1966) 207-222. MR 37, 1752.

48. Bazilevic, N. I. Almost periodic solutions of linear differential equations of infinite order with constant coefficients. (Russian). Maskov Pblast. Ped. Inst. Uien Zap. 166(1966) 275-289. MR 35, 4538.

49. ______, Lineare Differentialgleichungen unendlicher Ordnung mit konstanten Koeffizienten. Existenzsätze fur fastperiodschen Lösungen. Izvest. vyss ucebn Zaved. Mat. 91(1969) 9-19 (Russian). Z 195, 103.

50. Bekbaev, S., and Jataev, M. The investigation of certain critical cases of the stability of almost periodic motions. (Russian). Izv. Akal. Nauk Kazah SSR Ser. Fsy. Math. Nauk 1967 no. 3, 8-11. MR 35, 5110.

51. Bellman, R. Stability theory of differential equations. McGraw-Hill, N. Y. 1953.

52. Belova. Bounded solutions of systems of second order ordinary differential equations. Dokl. Akad. Nauk SSSR 180(1968) 266-268 appears in Soviet Math Dokl. A9(1968) 603-606. MR 39, 4482.

53. Berezanskiǐ, Y. M. On generalized almost periodic functions and sequences, related with the difference-differential equations. (Russian). Mat. Sb. 32(1953) 157-194. MR 14, 742.

54. Berg, I. D. On functions with almost periodic or almost automorphic first differences. J. Math. Mech. 19(1969) 239-246. MR 14, 746, 5.12.

55. Berman, D. L. Linear trigonometric polynomial operations in spaces of almost periodic functions. (Russian). Mat. Sb. (N. S.), 49(91)(1959) 267-280. MR 22, 3942.

56. Besicovitch, A. S. Almost periodic functions. Cambridge University Press, 1932.

57. Birjuk, G. I. On a theorem concerning the existence of almost periodic solutions for certain non linear differential systems with a small parameter. (Russian). Doklady Akad. Nauk SSSR 96(1954) 5-7.

58. ______, On the problem of the existence of almost periodic solutions for systems with a small parameter in a singular case. (Russian). Doklady Akad. Nauk SSSR 97(1954) 577-579.

59. Biroli, M. Solutions presque-périodiques des inéquations d'évolution paraboliques. Ann. Mat. Pura Appl. 88(1971) 51-70.

60. ______, Sur les solutions bornées et presque-périodiques des équations et inéquations d'évolution. Ann. Mat. Pura Appl. 93(1972) 1-79.

61. ______, Sur la solution bornée ou presque-périodique d'unéquation d'évolution du deuxieme ordre et du type elliptique. Boll U. M. I. 6(1972) 229-241. Z 244, 35010.

62. Blinov, I. N. Analytic representation of the solution of a system of linear differential equations with almost periodic coefficients which depend on a parameter. (Russian). Differencial'nye Uravnenija 1(1965) 1042-1053. MR 33, 4383, Z 196, 349.

63. ______, The reduction of a class of systems with almost periodic coefficients. (Russian). Mat. Zametki 2(1967) 395-400. MR 36, 6678.

64. ______, A certain class of improper systems. Differencial'nye Uravnenija 4(1968) 949-951. MR 38, 3493.

65. ______, Sulla esistenza ed unicita della soluzione limitata e della soluzione quasi periodica per una equazione parabolica con termine dissipativo non lineare discontinuo. Ric. di Mat. 19(1970) 93-110.

66. ______, An analytic solution of a linear system of differential equations with periodic coefficients depending upon a parameter. Differential Equations 1(1965) 679-691 (1967). Z 238, 34015.

67. ______, A class of irregular systems. Differential Equations 4(1968) 494-495 (1972). Z 235, 34019.

68. ______, Analytical representation of the solution of a system of linear differential equations with almost-periodic coefficients depending on a parameter. Differential Equations 1(1965) 812-821 (1967). Z 238, 34014.

69. ______, The loss of the property of reducibility by systems of linear differential equations with almost periodic coefficients. Math Notes 8(1970) 534-537 (1970). Z 215, 150, Z 224, 34032.

70. ______, The regularity of a class of linear systems with almost periodic coefficients. Differential Equations 3(1967) 764-768 (1971). Z 235, 34018.

71. Bochner, S. Über gewisse Differential' und allgemeinere Gleichungen deren Lösungen fastperiodisch sind. I. Teil. Der Existenzatz. Math. Ann. 102(1929) 489-504. FD 55,863.

72. _______, Über gewisse Differential' und allgemeinere Gleichungen deren Lösungen fastperiod sind. II. Teil. Der Beschränktheitssatz. Math. Ann. 103(1929) 588-597. FD 56, 1043.

73. _______, Über gewisse Differential' und allgemeinere Gleichungen deren Lösungen fastperiodisch sind. III. Teil. Systeme von Gleichungen. Math. Ann. 104(1931) 579-587. Z 1, 275.

74. _______, Allgemeine lineare Differenzengleichungen mit asymptotisch konstanten Koeffizienten. Math. Z. 33(1931) 426-250. Zbl 1, 67.

75. _______, Remark on the integration of almost periodic functions. J. London Math. Soc. 8(1933) 250-254. Z 8, 12.

76. _______, Homogeneous systems of differential equations with almost periodic coefficients. J. London Math. Soc. 8(1933) 283-288. Z 7, 347.

77. _______, Fastperiodische Lösungen der Wellengleichung. Acta Math. 62(1934) 227-237. Z 9, 163.

78. _______, Almost periodic solutions of the inhomogeneous wave equations. Proc. Nat. Acad. Sci. U. S. 46(1960) 1233-1236. MR 22, 9726.

79. _______, A new approach to almost periodicity. Proc. Nat. Acad. Sci. U. S. 48(1962) 2039-2043. MR 26, 2816, 1.14, 5.12, 6.10.

80. Bochner, S., and Neumann, J. von. On compact solutions of operational-differential equations. I. Ann. Math. 36(1935) 255-290. Z 11, 20.

81. Bogatyrev, B. On the question of the construction of solutions of linear differential equations with quasi-periodic coefficients in a Banach space. (Russian). Trudy Sem. Mat. Fiz. Nelinein. Koleban. i(1968) vyp. 2, 78-87. MR 44, 580.

82. Bogdanowicz, W. M., and Welch, J. N. On a linear operator connected with almost periodic solutions of linear differential equations in Banach spaces. Math. Ann. 172(1967) 327-335. MR 35, 5711.

83. Bogdanowicz, W. M. On the existence of almost periodic solutions for systems of ordinary differential equations in Banach spaces. Arch. Rational Mech. Anal. 13(1963) 364-370. MR 27, 1675, 8.6.

84. Bogoliubov, N. Quelques theoremes sur la stabilité des mouvements. Enseignement Math. 34(1936) 337-346. Zbl. 14, 259.

85. _______, Quasiperiodic solutions in problems of nonlinear mechanics. (Russian). First Math Summer School, Part I pp. 11-101. Izdat. "Naukova Dumka", Kiev, 1964. MR 35, 468.

86. Bogoliubov N., and Kryloff, N. New methods in nonlinear mechanics. (Russian). Moscow. 1934.

87. _______, Sur les solutions quasi-périodiques des équations de la mécanique nonlinéaire. C. R. Acad. Sci. Paris, 199(1934) 1592-1593. Z 11, 67.

88. Bogoliubov, N., and Mitropolskii, Y. A. Asymptotic methods in the theory of nonlinear vibrations. (Russian). Moscow, 1955.

89. _______, Regimes quasi-periodiques dans les systems oscillants non lineaires. Colloq. Internat. Centre nat Rech. Sci. 148(1965) 181-192. (Russian). French translation 193-204. Z 196, 370.

90. Bohl, P. Ueber eine Differentialgleichung der Störungstheorie. J. f. reine u. angew. Math. 131(1906) 268-321.

91. Bohr, H. Fastperiodische Funktionen. Springer-Verlag, Berlin, 1932.

92. _______, Stabilitet og Naestenperiodicitet. Mat. Tidsskrift (1933) 21-25. (Collected Works VII).

93. _______, Über fastperiodische Bewegungen. Neuvième Congrès des Mathematiciens Scandinaves. Helsingfors, 1938 39-61. (Collected Works VII).

94. Bohr, H., and Jessen, B. Über fastperiodische Bewegungen auf einem Kreise. Annali Scuola Norm. Sup. Pisa, Scienze fisiche e matimatiche, 1(1932) 385-398. (Collected Works VII).

95. _______, Mean motions and almost periodic functions. Colloques Internationaux du C. N. R. S., Analyse Harmonique, Paris, 1949 75-84. (Collected Works VIII).

96. Bohr, H., and Neugebauer, O. Über lineare Differentialgleichungen mit konstanten Koeffizienten und fastperiodischer rechter Seite. Nachr. Ges. Wiss. Göttingen, Math.-Phys. Klasse (1926) 8-22. (Collected Works VII), 5.12.

97. Borhuhov, L. A linear integral equation with almost periodic kernel and free term. (Russian). Doklady Akad. Nauk SSSR 57(1947) 647-649. MR 9,356.

98. _______, On periodic and almost periodic solutions of the equations y' + q(x)y = f(x). (Russian). Nauc. Ez. Saratovsk. Univ. (1954) 656-657.

99. _______, On almost periodicity of solutions of some linear differential systems with almost periodic coefficients. (Russian). Nauc. Ez. Saratovsk. Univ. (1954) 659-660.

100. _______, On almost periodic solutions of certain differential equations. (Russian). Nauc. Ez. Saratovsk. Univ. (1954) 660-661.

101. _______, On periodicity and almost periodicity of solutions of certain second order linear differential equations. (Russian). Nauc. Ez. Saratovsk. Univ. (1954) 661-663.

102. Brauers, N. Différentiation et intégration des fonctions presque-périodiques de plusieurs variables réelles. Acta Univ. Latviensis (1939) 235-263. MR 1, 330.

103. _______, Sur l'integration des fonctions presque-périodiques de deux variables. Comm. Math. Helvetici 11(1939) 330-335. Z 21, 222.

104. Brommundt, E. Approximate solutions of quasiperiodic differential equation. J. Math. Anal. Appl. 30(1970) 252-263. MR 41, 576.

105. Burd, V. S. On the branching of almost periodic solutions of nonlinear ordinary differential equations. (Russian). Doklady Akad. Nauk SSSR. 159(1961) 239-242. MR 31, 2457.

106. _______, Dependence on a parameter of almost periodic solutions of differential equations with a deviating argument. (Russian). Akad. Nauk Azerbaidzan SSR Dokl. 21(1965) 3-7. MR 32, 4348.

107. Burd, V. S., and Baberov, T. The stability of braching periodic solutions of certain systems of differential equations. (Russian). Dokl. Akad. Nauk SSSR 176(1967) 991-993. MR 36, 5444.

108. Burd, V. S.; Kolesov, J.; and Krasnosel'skii. On the existence and constancy of sign of Green's function of scalar equations of high order with almost periodic coefficients. Math of USSR-Izvestia (translation) 3(1969) 1319-1334. Z 206, 96.

109. _______, The bifurcation of the almost periodic solutions of singularly perturbed differential equations. (Russian). UspehiMat. Nauk 25(1970) no. 1(151) 189-190. MR 41, 7226.

110. Burd, V. S., and Kolesov, J. On the dichotomy of solutions of functional-differential equations with almost periodic coefficients. Soviet Math 11(1970) 1650-1653. MR 42, 8040.

111. Burd, V. S.; Zabreiko, P. P.; Kolesov, J.; and Krasnosel'skii M. A. The averaging principle and bifurcation of almost periodic solutions. (Russian). Dokl. Akad. Nauk SSSR 187(1969) 1219-1221. MR 41, 7211.

112. Burnat, M. Die Spektralclarstellungen eineger differentialoperatorem mit periodischer Koeffizienten im Raume der fastperiodischen Funktionen. Studia Math. 25(1964-65) 33-64. MR 31, 6018.

113. Burton, T. A. Linear differential equations with periodic coefficients. Proc. Amer. Math. Soc. 17(1966) 327-329 MR 32, 7855.

114. Bylov, B. F. On stability from above of the greatest characteristic index of a system of linear differential equations with almost-periodic coefficients. (Russian). Mat. Sb. (N. S.) 48(90)(1959) 117-128. MR 22, 1710.

115. _______, The structure of the solutions of a system of linear differential equations with almost periodic coefficients. (Russian). Mat. Sb. (N. S.) 66(108)(1965) 215-229. MR 31, 2450.

116. _______, Almost reducible systems. (Russian). Sibirsk. Mat. Z. 7(1966) 751-784. MR 34, 1588.

117. Bystrenin, V. V. On almost periodic solutions of certain ordinary differential equation. (Russian). Doklady Akad. Nauk SSSR 33(1941) 387-389. MR 5, 120.

118. Cameron, R. Linear differential equations with almost periodic coefficients. Duke math. J. 1(1935) 356-360. Zbl. 12, 297.

119. _______, Linear differential equations with almost periodic coefficients. Ann. Math. 37(1936) 29-42. Z 14, 7.

120. _______, Almost periodic properties of bounded solutions of linear differential equations with almost periodic coefficients. J. Math. Phys. 15(1936) 73-81. Z 13, 263.

121. _______, Linear differential equations with almost periodic coefficients. Acta Math. 6(1938) 21-56. Z 18, 211, 7.7.

122. _______, Quadratures involving trigomometric sums. J. Math. Phys. 19(1940) 161-166. MR 2, 93, 6.10.

123. Caracosta, G., and Doss, R. Sur l'intégrale d'une fonction presque-périodique. C. R. Acad. Sci. Paris 246(1958) 3207-3208. MR 20, 2574.

124. Carroll, F. W. On bounded functions with almost periodic differences. Proc. Amer. Math. Soc. 15(1964) 241-243. MR 28, 5137.

125. Cartwright, M. L. Almost periodic solutions of certain second order differential equations. Rend. Sem. Mat. Fis. Milano 31(1961) 100-110. MR 27, 4989.

126. _______, Almost periodic solutions of equations with periodic coefficients. Non-linear problems (Proc.-Sympo. Madison, Wisc. 1962) 207-218. Univ. of Wisc. Press. MR 32, 4326.

127. _______, Almost periodic solutions of systems of two periodic equations. (Russian summary). Analytic methods in the theory of non-linear vibrations. (Proc. Internat. Sympos, Non-linear Vibrations, Vol. 1, 1961) pp. 256-263. Izdat. Akad. Nauk Ukrain. SSR, Keiv 1963. MR 28, 600.

128. _______, Almost periodic flows and solutions of differential equations. Math. Nachr. 32(1966) 257-261. MR 37, 531.

129. _______, Almost periodic flows and solutions of differential equations. Proc. London Math. Soc. (3) 17(1967) 768. Corrigenda (3) 17(1967) 768. MR 36, 6697.

130. _______, Almost periodic differential equations and almost periodic flows. J. Differential Equations 5(1969) 167-181. MR 39, 548.

131. _______, Comparison theorems for almost periodic functions. J. London Math. Soc. (2) 1(1969) 11-19. MR 44, 5710, 4.9.

132. _______, Almost periodic solutions of differential equations and flows. Global differentiable Dynamics, Proc. Conf. Case Western Reserve Univ., Cleveland, Ohio 1969. Lecture Notes Math. 235 35-43 (1971). Z 238, 34078.

133. Castellana-Rizzonelli, P. Sulle funzioni limitate, a valori vettoriali, con differenze quasi periodiche. Tamburini Editore, Milano 1963 p. 14.

134. Čeresiz, V. M. On almost periodic solutions of nonlinear systems. (Russian). Doklady Akad. Nauk SSSR 165(1965) 281-284. MR 33, 7635.

135. ______, Almost-periodicity of bounded solutions of nonlinear systems. (Russian). Dokl. Akad. Nauk SSSR 173(1967) 275-277. MR 35, 3152.

136. ______, On the stability of almost periodic solutions. Soviet Math 13(1972) 397-399.

137. Cerneau, S. Sur la construction de solutions presque-périodiques de problèmes de perturbation singulière. C. R. Acad. Sci. Paris 260(1965) 768-771. MR 30, 2197.

138. Chang, K. W. Almost periodic solutions of singularly perturbed systems of Differential Equation. J. Differential Equations. 4(1968) 300-307. MR 37, 511.

139. ______, Funzioni quasi-periodiche ed equazioni differenziali. Rend. Sem. Mat. Torino 11(1951-1952) 47-74.

140. ______, Sopra qualche concetto della teoria delle funzioni quasi-periodiche. Matematiche (Catania) 12(1957) 1-17.

141. Chow, S. Perturbing Almost periodic Diff. Eq. Bull. Amer. Math. Soc. 1970(76) 421-424. MR 40, 4547.

142. Conley, C. C, and Miller, R. K. Asymptotic stability without uniform stability. J. Differential Equations 1(1965) 333-336. MR 34, 1619, 11.11.

143. Cooke, R. Almost periodicity of bounded and compact solutions of differential equations. Duke Math Journal 36(1969) 273-276. Z 176, 91.

144. Coppel, W. A. Dichotomies and reducibility. J. Differential Equations, 4(1967) 500-521. MR 36, 6699, 7.8.

145. ______, Almost periodic properties of ordinary differential equations. Ann. di Mat. Pura ed appl. 76(1967) 27-50. MR 36, 4076, 4.9, 5.12, 6.10, 8.6, 11.11.

146. ______, Dichotomies and reducibility. II. J. Differential Equations 4(1968) 386-398. MR 37, 4338, 7.8.

147. Corduneanu, C. Solutions presque-périodiques des equations differentielles non lineaires du second ordre. Com. Acad. R. P. Romine 5 793-797 (1955). MR 17, 39.

148. _______, Solutii aproape-periodice ale ecuatiilor diferentiale neliniare de ordinul al doilea. Comunicările Acad. R. P. R. 5(1955) 21-26.

149. _______, Soluaii asimptotic aproape-periodice ale ecuatiilor diferentiale neliniare de ordinul al doilea. Studii si cercetări stiintifice. Matematică (Iasi) 6(1955) 1-4. MR 20, 4677.

150. _______, Cîteva consideratii în legătură cu unele sisteme neliniare de ecuatii diferentiale. Studii şi cerc. ştiint. Mathematica (Iasi) VII(1956) fasc. 2 13-23.

151. _______, Solutions presque-périodiques de certaines equations paraboliques. Mathematica (Cluj) 9 (32)(1967) 241-244. MR 41, 623.

152. Cronin, J. Almost-periodic solutions and critical roots. Duke Math. J. 29(1962) 663-669. MR 25, 5235.

153. Demidovich, B. P. On a case of almost periodicity of solutions of an ordinary differential equation of first order. (Russian). Usp. Mat. Nauk 8 f.6(1953) 103-106. MR 15, 812.

154. _______, On bounded solutions of a certain nonlinear system of ordinary differential equations. (Russian). Mat. Sb. 40(82)(1956) 73-94. MR 18, 738.

155. _______, On bounded solutions for some quasi-linear systems. Doklady Akad. Nauk SSSR 138(1961) 1273-1275. MR 24, 458.

156. Deysach, L. G. and Sell, G. R. On the existence of almost periodic motions. Michigan Math. J. 12(1965) 87-95. MR 30, 3279, 11.11.

157. Denjoy, A. Sur les trajectoires du tore. C. R. Acad. Sci. Paris 251(1960) 175-177. MR 22, 8166.

158. Diliberto, S. An application of periodic surfaces. Cont. to the Theory of Non-Linear Oscillations. S. Lefschetz, ed. Princeton U. Press (1956) 257-261. MR 19, 144.

159. Doss, R. On bounded functions with almost periodic differences. Proc. Am. Math. Soc. 12(1961) 488-489. MR 23, A 3424, 5.12.

160. _______, On the almost-periodic solutions of a class of integro-differential-difference equations. Ann. Math. (2) 81(1965) 117-123. MR 30, 408, 5.12.

161. Emzarov K., and Tulegenov M. An existence theorem for almost periodic solutions of a differential equation with a small parameter in a Banach space. (Russian). Vestnik Akad. Nauk Kazah SSR 22(1966) no. 10 (258) 42-44. MR 34, 2990.

162. Erugin, N. P. Linear systems of ordinary differential equations with periodic and quasi-periodic coefficients. Izdat. Akad. Nauk USSR, Minsk 1963. 272 pp. 1.05r.

163. _______, Linear systems of ordinary differential equations with periodic and quasi-periodic coefficients. Academic Press (1966). MR 34, 6175.

164. Esçlangon, E. Sur les intégrales quasi-périodiques d'équations différentielles linéaires. C. R. Acad. Sci. Paris 158(1914) 1254-1256.

165. _______, Sur les integrales bornees d'une equation differentielle linéaire. C. R. Acad. Sci. Paris 160(1915) 475-478.

166. _______, Sur les integrales quasi-périodiques d'une equation differentielle. C. R. Acad. Sci. Paris 160(1915) 652-653, 488-489.

167. Ezeilo, J. O. C. On the existence of an almost periodic solution of a non-linear system of differential equations. Contributions to Differential Equations 3(1964) 337-349. MR 30, 2187.

168. _______, A generalization of a result of Demidovic on the existence of a limiting regime of a system of differential equation. Portugal. Math. 24(1965) 65-82. MR 34, 4616.

169. _______, On the existence of almost periodic solutions of some dissapative second order differential equations. Ann. Mat. Pura Appl. 74(1966) 399. MR 34, 4609.

170. Fan, K. Les fonctions asymptotiquement presque-périodiques d'une variable entière et leur application a l'étude de l'itération des transformations continues. Math. Z. 48(1942/43) 685-711. MR 5, 99.

171. Favard, J. Sur les équations différentielles a coefficients presque-périodiques. Acta Math. 51(1927) 31-81. Fdm 53, 409.

172. _______, Lecons sur les fonctions presque-périodiques. Gauthier-Villars, Paris 1933. Z 7, 343, 4.9.

173. _______, Sur certains systèmes différentiels scalaires linéaires et homogènes à coefficients presque-périodiques. Ann. Mat. Pura ed Appl. (4) 61(1963) 297-316. MR 27, 4990.

174. _______, Sur les équations différentielles scalaires presque-périodiques. J. Math. pures appl. (9) 43(1964) 87-97. MR 29, 1388.

175. Fenchel, W. Om Bevaegelser, der er naestenperiodiske paaner Flytninger. Mat. Tidsskrift B (1937) 75-80. Z 16, 356.

176. Fenchel, W. and Jessen, B. Über fastperidische Bewegungen in ebenen Bereichen und auf Flächen. Mat.-fysiske Medd. 13 6(1935) 1-18. Z 11, 346.

177. Fink, A. M. Almost automorphic and almost periodic solutions which minimuzc functionals. Tohoku Math. J. 20(1968) 323-332. MR 39, 549.

178. _______, Uniqueness theorems and almost periodic solutions to second order differential equations. J. Diff. Eqs. 4(1968) 543-548. MR 38, 369.

179. _______, Compact families of almost periodic functions. Bull. Amer. Math. Soc. 75(1969) 770-771. MR 40, 640.

180. _______, Compact families of almost periodic functions and an application of the Schauder fixed point theorem. SIAM J. Appl. Math. 17(1969) 1258-1262. MR 41, 2249, 8.6.

181. _______, Almost periodic solutions with module containment. Advances in Differential and Integral Equations. SIAM (1969) 177.

182. _______, Extensions of almost automorphic sequcnces. J. Math. Anal. Appl. 27(1969) 519-523. MR 39, 6017.

183. _______, Almost periodicity of the inverse of a fundamental matrix. Proc. Amer. Math. Soc. 27(1971) 527-529. MR 42, 4822.

184. _______, A nonresonance condition for a.p. solutions to difference equations. J. Math. Anal. Appl. 33(1971) 306-309. Z 214, 99.

185. _______, Almost periodic functions invented for specific purposes. SIAM Review 14(1972) 572-581.

186. _______, Semi-separated conditions for almost periodic solutions. J. Diff. Eqs. 11(1972) 245-251. MR 45, 3855, 9.7, 10.5, 11.11.

187. Fink, A. M., and Frederickson, P. Ultimate boundedness does not imply almost periodicity. J. Diff. Eqs. 9(1971) 280-284. MR 42, 8007.

188. Fink, A. M., and Seifert, G. Liapunov Functions and almost periodic solutions for almost periodic systems. J. Diff. Eqs. 5(1969) 307-313. MR 38, 3517, 11.11.

189. ______, A nonresonance condition for almost periodic solutions of ordinary differential equations. Ann. Mat. Pura Appl. 81(1969) 69-76. MR 39, 7211.

190. ______, Non-resonance conditions for the existence of almost periodic solutions of almost periodic systems. SIAM J. Appl. Math. 21(1971) 362-366, 8.6.

191. Foias, C. Essais dans l'étude des solutions des équations de Navier-Stokes dans l'espace. L'unicité et la presque-périodicité des solutions "petites". Rend. Sem. Mat. Univ. Padova 32(1962) 261-264. MR 26, 5311.

192. Foias, C., and Zaidman, S. Almost periodic-solutions of parabolic systems. Ann. Scuola Norm. Sup. Pisa (3) 15(1961) 247-262. MR 24, A 3348.

193. Fomin, V. N. The dynamic instability of linear systems with almost periodic coefficients. (Russian). Dokl. Akad. Nauk SSSR 178(1968) 43-46. MR 37, 1717, Z 188, 399.

194. ______, The resonance of oscillations of linear systems acted on by an almost periodic parametric perturbation. (Russian). Problems of Math. Anal. No. 2: Linear Operators and Operator Equations. (Russian). pp. 28-79. Izdat. Leningrad. Univ. Leningrad, 1969. MR 45, 688.

195. Franklin, P. Almost period recurrent motions. Math. Z. 30(1929) 325-331. Fdm 55, 1094, 11.11.

196. Frederickson, P. O. Generalized series solution of an almost periodic differential equation. J. Diff. Eqs. 2(1966) 243-264. MR 34, 7882.

197. ______, Dirichlet series solutions for certain functional differential equations. Lecture Notes 243(1971) Japan-United States Seminar on Ordinary Differential and Functional Equations 249-254. Z 233, 34075.

198. ______, Global solutions to certain nonlinear functional differential equation. J. Math. Anal. Appl. 33(1971) 355-358. MR 42, 3380.

199. Frederickson, P. O., and Lazer, A. C. Necessary and Sufficient Damping in a Second-Order Oscillator. J. Diff. Eqs. 5(1969) 262-270. MR 38, 2380.

200. Friedman, M. Quasi-periodic solutions of nonlinear ordinary differential equations with small damping. Bull. Amer. Math. Soc. 73(1967) 460-464. MR 37, 5477.

201. Gelman, A. E. The reducibility of a certain class of simultaneous differential equations containing quasi-periodic (a.p.) coefficients. Dokl. Akad. Nauk SSSR (N.S.) 116(1951) 535-537. MR 20, 1017.

202. ______. On the reducibility of systems with a quasiperiodic matrix. (Russian). Differencial'nye Uravnenija 1(1965) 283-294. MR 34, 7844, Z 196, 372.

203. Gheorghiu, N. Asupra solutiilor aproape-periodice si asimptotic aproape-periodice ale ecuatiilor diferentiale neliniare di primul ordin. Analele st. Univ. Iasi (s.n.) 1(1955) 17-20. MR 18, 899.

204. Golomb, M. Expansions and boundedness theorems for solutions of linear differential systems with periodic or almost periodic coefficients. Archiv for Rat. Mech. and Analysis. 2(1958) 284-308. MR 21, 3624.

205. ______, Solution of certain nonautonomous differential systems by series of exponential functions. Ill. Jour. Math. 3(1959) 45-65. MR 21, 3625.

206. ______, On the reducibility of certain linear differential systems. Journal fur die Reine und Angw. Math. 205(1961) 171-185. MR 23, A 1872.

207. Günzler, H. Fastperiodische Lösungen linearer hyperbolischer Differentialgleichungen. Math. Z. 71(1959) 223-250. MR 21, 7403.

208. ______, Hyperbolic differential equations and almost periodic functions. Rend. Sem. Mat. Fis. Milano 34(1964) 165-201. MR 30, 5057.

209. ______, Beschränktheitseigenschaften für Lösungen nichtlinearer Wellengleichungen. Math. Ann. 167(1966) 75-104. MR 34, 3111.

210. ______, Integration of almost-periodic functions. Math. Z. 102(1967) 253-287. MR 36, 3066.

211. Günzler, H., and Zaidman, S. Abstract almost periodic differential equations, Abstract spaces and Approximation. 387-392. Birkhaüser, Basel 1969. MR 41, 3946.

212. ______, Almost periodic solutions of a certain abstract differencial equation. J. Math. Mech. 19(1969) 155-158. MR 40, 7607.

213. Gur'janov, A. On sufficient conditions for the regularity of second order systems of linear ordinary differential equations with uniform asymptotically almost periodic coefficients. Vestnik Leningrad Univ. 25(1970) 23-27. MR 43, 2273.

214. Gurtonvik. A. S., and Neimark J. J. On the question of the stability of quasi periodic motions. Diffencialnje Uravenija 5(1969) 824-832. MR 40, 2967.

215. Hakimzanova, S. I. An application of operatonal calculus to the investigation of quasiperiodic processes. (Russian). Proc. Second Kazakhstan Interuniv. Sci. Cong. Math. Mech. (1965) (Russian). pp. 102-104. Izdat. "Nauka" Kazah. SSR Alma-Ata, 1968. MR 41, 578.

216. Halanay A. Solutii aproape-periodice ale ecuatiei Riccati. Studii si cercetări matematice. 4(1953) 245-354. MR 16, 475.

217. ______, Solutii aproape-periodice ale unor sisteme neliniare. Gazeta mat. si fiz. A 7(1955) 396-399. MR 18, 395.

218. ______, Solutii aproape-periodice pentru sistemele de ecuatii diferentiale neliniare. Comunicările Acad. R. P. R. 6(1956) 13-17. MR 18, 212.

219. ______, Intoducere in teoria calitativă a ecuatillor diferentiale. Bucuresti, 1956.

220. ______, Solutions presque-périodiques des systemes d'équations differentielles à argument retardé contenant un petit paramètre. Comunicările Acad. R. P. R. 9(1959) 1237-1242. MR 6918.

221. ______, Periodic and almost-periodic solutions of systems of differential equations with lagging argument. (Russian). Rev. Math. pures et appl. 4(1959) 685-691. MR 22, 2774.

222. ______, Almost periodic-solutions of systems of differential equations with a lagging argument and small parameter. (Russian). Rev. Math. pures et appl. 5(1960) 75-79. MR 24, A 887.

223. _______, Solutions presque-périodiques des systèmes linéaires héréditaires. C. R. Acad. Sci. Paris 257(1963) 827-829. MR 27, 3880.

224. _______, Periodic and almost-periodic solutions of certain singularly perturbed systems with retarded argument. (Russian). Rev. Math. pures et appl. 8(1963) 285-292. MR 31, 2467.

225. _______, Almost periodic solutions of systems with a small parameter in a certain critical case. (Russian). Rev. Math. Pures Appl. (Bucharest) 8(1963) 397-403. MR 31, 1429.

226. _______, Teoria calitativă a ecuatillor diferentiale. Editura Acad. R. S. Romainia, Bucharest, 1963. (English translation, Academic Press, 1966). MR 35, 6938.

227. _______, Almost-periodic solutions of linear delay-systems. (Russian). Rev. Roumaine de Math. pures et appl. 9(1964) 71-79. MR 31, 2469.

228. _______, On the method of averaging for differential equations with retarded argument. J. Math. Anal. Appl. 14(1966) 70-76. MR 33, 1562.

229. _______, Quasi-periodic solution for linear systems with time lag. Rev. Raum. Math pur appl. 14(1969) 1463-1474, 1269-1276. MR 41, 3938, 3937.

230. Hale, J. K. Oscillations in nonlinear Systems. McGraw-Hill, New York, 1963.

231. _______, Periodic and almost-periodic solutions of functional differential equations. Arch. Rational Mech. Anal. 15(1964) 289-304. MR 1395.

232. Hale, J. K., and Seifert, G. Bounded and almost-periodic solutions of singularly perturbed equations. J. Math. Anal. Appl. 3(1961) 18-24. MR 25, 3229.

233. _______, Bounded and almost-periodic solutions of singularly perturbed equations. Qualitative methods in the theory of non-linear vibrations. Proc. Intern. Symp. Non-linear Vibrations, Kiev. vol. II, 1961 427-432. MR 28, 3200.

234. Harasahal, V. H. Almost-periodic solutions of non-linear systems of differential equations. (Russian). Prikl. Mat. Meh. 24(1960) 565-567. MR 23, A 1110.

235. _______, The structure of solutions and the correctness of linear systems of differential equations with quasi-periodic coefficients. (Russian). Doklady Akad. Nauk SSSR 146(1962) 1290-1293. MR 25, 5233.

236. ______, On quasi-periodic solutions of systems of ordinary differential equation. (Russian). Prikl. Mat. Mech. 27(1963) 672-682. MR 28, 3205.

237. ______, On quasi-periodic solutions of differential equations. (Russian). Izv. Vyss. Ucebn. Zaved. Matematika, No. 2(39) (1964) 152-164. MR 29, 311.

238. ______, The work of Alma-Ata mathematicians in the theory of almost periodic solutions, and certain problems of differential equations. (Russian). Proc. Second Kazakhstan Interuniv. Sci. Conf. Math. Mech. (1965) (Russian) pp. 46-50. Izdat. "Nauka" Kazah. SSR, Alma-Ata, 1968. MR 41, 2125.

239. ______, The regularity of linear systems of differential equations with quasi periodic coefficients. Certain problems in Diff. Eq. 3-6(1969). MR 40, 2968.

240. Hartman, P. Mean Motions and almost periodic functions. Trans. Am. Math. Soc. 46(1939) 66-81. MR 1, 12.

241. Hartman, P.; van Kampen, E. R.; and Wintner, A. Mean motions and distribution functions. Am. J. Math. 59(1937) 261-269.

242. Hartman, P., and Wintner, A. On the classical existence theorem of linear differential equations. Amer. J. Math. 71(1949) 859-864. MR 11, 101.

243. Hartman, S. Les integrales de fonctions presque-périodiques et les sections de series de Fourier. Studia Math. 22(1962/63) 147-160. MR 26, 550.

244. Hayashi, C.; Shibayama, H.; and Ueda, Y. Quasi-periodic oscillations in a self-oscillatory system with external force. Analytic methods in the theory of non-linear vibrations. Proc. Intern. Symp. non-linear Vibrations, vol. 1, 1961 495-509. MR 28, 2309.

245. Hemp, G., and Sethna, P. On dynamical systems with high frequency parametric excitation. Int. J. Mon. Cinker Mech. 3(1968) 351-365. MR 39, 3097.

246. Hermes, H. A survey of recent results in differential equations. SIAM Review 15(1973) 453-468.

247. Hurt, J. J., and Schaeffer, A. J. Critical linear difference equation: a study in pathology. J. Math. Anal. Appl. 33(1971) 408-424 , 5.12.

248. Husson, E L'aire couverte par une trajectoire dynamique; la presque-periodicite de la trajectoire. J. Math. pures appl. IX. s. 16(1937) 101-104. Zbl. 16, 86.

249. Ivanov, V. N. On linear differential operators in the space of almost-periodic functions. (Russian). Trudy Saratovsk. Inst. Mch. Selsk., fasc. 38(1965) 141-149.

250. Jakubovic, V. A. The method of matrix inequalities in the theory of stability of nonlinear control systems. I. (Russian). Avtomat. i Telemeh. 25(1964) 1017-1029. MR 29, 4953.

251. ______, Periodic and almost-periodic limit regimes of automatic control systems with some discontinuous non-linearities. (Russian). Doklady Akad. Nauk SSSR 171(1966) 533-537. MR 34, 6237.

252. Jessen, B. Remark on the theorems of R. Petersen and S. Takahashi. Mat. Tidsskrift (1935) 85-86. Z 13, 112.

253. ______, Über die Sekularkonstanten einer fastperiodischen Funktion. Math. Ann. 111(1935) 355-363. Z 12, 66.

254. ______, Om Sekularkonstanten for en naestenperiodisk Funktion. Mat. Tidsskrift B (1937) 45-48. Z 16, 356.

255. ______, Mouvement moyen et distribution des valeurs des fonctions presque-périodiques. Comptes Rendus du dixieme Congres des Math. Scandinaves, Kobenhavn, 1946 301-312. MR 8, 578.

256. ______, Mean motion and almost periodic functions. Proc. Second Canadian Math. Congress, Vancouver, 1949 76-92. MR 13, 342.

257. Jessen, B., and Tornehave, H. Mean motions and zeros of almost-periodic functions. Acta. Math. 77(1945) 137-279. MR 7, 438.

258. Kamenskii, G. A. Smoothing dynamical systems and almost periodic solutions of nonlinear differential equations with time lag. Kacestv. Met. Teor. nelin. Koleb. 2. Trudy 5 mezdunarod. Konf. 1969 200-204. (Russisch. Engl. Zusammenfassung)(1970). Z 235, 34139.

259. Kapisev, K. K. Quasiperiodic solutions of nonlinear systems of differential equations containing a small parameter. (Russian). Vestnik Akad. Nauk Kazah. SSR 22(1966) no. 6 (254) 42-47. MR 33, 6031.

260. Kapisev, K. K., and Harashal, V. H. Quasiperiodic solutions of systems of differential equations containing a small parameter. (Russian). Izv. Akad. Nauk Kazah. SSR Ser. Fiz. Mat. Nauk (1966) no. 3, 76-82. MR 35, 438.

261. Kato, J. Uniform Asymptotic Stability and Total Stability. Tohuku Math. J. 22(1970) 254-269. MR 42, 3381, 11.11.

262. Kato, J., and Yoshizawa, T. A relationship between uniformly asymptotic stability and total stability. Funkcial. Ekvac 12(1969/70) 233-238. MR 41, 2147.

263. Kempner, G. A. Almost periodic functional differential equations. SIAM J. Appl. Math. 16(1968) 155-161. MR 36, 5466.

264. Kitagawa, T. The application of the theory of Cauchy's series on the solutions of the operational-difference equation. Tohoku Math. J. 44(1937) 139-141. Zbl 18, 366.

265. Kneser, H. Periodische Differentialgleichungen und fastperiodische. Funktionen. Ann. Mat. pura appl., IV s. 11(1933) 181-185. Zbl 6, 304.

266. Kokotov, M. I. The almost periodicity of the solutions of the equation $\partial^2u/\partial t^2 = \Delta u - q(x)u + f(x,t)$. (Russian). Dokl. Akad. Nauk SSSR 197(1971) 1014-1017. MR 43, 7757.

267. Konovalov, J. Über fastperiodische Lösungen quasiharmonischer Systems mit Retardierung. Izvest. vyšs ŭcebn. Zaved. Mat. 89(1969) 62-69. MR 42, 2127.

268. Kovanko, A. S., and Lisevič, L. M. Some properties of the indefinite integral and derivative of an S_p almost-periodic function. (Ukr.). Dopovidi Akad. Nauk Ukr. R. S. R. (1963) 705-706. MR 29, 2601.

269. ______, Almost-periodic solutions of certain differential equations with almost-periodic right sides. (Ukr). Visnik Lwow. Univ. Ser. Math. (1965) fasc. 2, 3-8. MR 35, 869.

270. Krasinskii, G. A. Parametric resonance in canonical systems of linear differential equations with quasiperiodic coefficients. (Russian). Dokl. Akad. Nauk SSSR 180(1968) 526-529. MR 38, 374.

271. Krasnosel'skii, M. A.; Burd, V. S.; and Kolesov, J. S. (Russian). Nonlinear almost periodic oscillations. Izdat. "Nauka", Moscow, 1970 351 pp. 1.33 r. MR 45, 7183.

272. _______, Nonlinear Almost Periodic Oscillations (1970) 352 pages. (Russian). Z 213, 107.

273. Krasnoselskii, M. A., and Perov, A. I. A principle concerning the existence of bounded, periodic and almost-periodic solutions for systems of ordinary differential equations. (Russian). Doklady Akad. Nauk SSSR 123(1958) 235-238. MR 21, 3621.

274. Krein, M. G. Lectures on stability theory of solutions of differential equations in Banach space. (Russian). Kiev. 1964, 186 p.

275. Kryloff, N. and Bogoliuboff, N. Quelques examples d'oscillations non linéaires. C. R. Acad. Sci. Paris 194(1932) 957-960. Zbl. 4, 10.

276. Kudakova, R. V. Quasiperiodic solutions of non-linear systems of differential equations containing a small parameter. (Russian). Izv. Akad. Nauk Kazah. SSR Ser. Fiz. Mat. Nauk 1963, 48-52. MR 28, 4214.

277. Kurzweil, J, and Vejvoda, O. On the periodic and almost-periodic solutions of a system of ordinary differential equations. Czechoslavak Math. J. 5(80)(1955) 362-372. (Russian, English summary). MR 17, 849.

278. Ladyženskaja, O. A. The mixed problem for the hyperbolic equations. (Russian). Moscow, 1953. MR 17, 160.

279. Lakshmikantham, and Leela, S. Almost periodic systems and differential inequalities. Proceedings United States-Japan Seminar on Differential and Functional Equations. (1967) 549-554. MR 39, 5883.

280. _______, Almost periodic systems and differential inequalities. J. Math. Phys. Sci. Madras 1(1967) 286-301. Z 195, 102.

281. Langenhop, C. E., and Seifert, G. Almost-periodic solutions of second order non-linear differential equations with almost periodic forcing. Proc. Am. Math. Soc. 10(1959) 425-432. MR 21, 3623.

282. Langenhop, C. E. Note on almost periodic solutions of non-linear differential equations. J. Math. Phys. 38(1959/60) 126-129. MR 21, 6459.

283. _______, On bounded matricies and kinematic similarity. Trans. Am. Math. Soc. 97(1960) 317-326. MR 22, 5788, 7.7.

284. Lazer, A. C. Characteristic exponents and diagonally dominant linear differential systems. J. Math. Anal. and Appl. 35(1971) 215-229. MR 43, 6537, 7.8.

285. Levitan, B. M. On linear differential equations with almost-periodic coefficients. (Russian). Doklady Akad. Nauk SSSR 17(1937) 285-286.

286. ______, On an integral equation with an almost periodic solution. Bull. Amer. Math. Soc. 43(1937) 677-679. Zbl. 17, 357.

287. ______, Über lineare Differentialgleichungen mit fast-periodischen Koeffizienten. C. R. Acad. Sci. URSS, N. s. 17(1937) 289-290. Zbl. 18, 59.

288. ______, The application of generalized displacement operators to linear differential equations of the second order. Uspehi Matem. Nauk (N. S.) 4, no. 1 (29)(1949) 3-112. (Russian). MR 11, 116.

289. ______, Almost-periodic functions. (Russian). Moscow, 1953, 4.9.

290. Lieberman, B. Quasi-periodic solutions of Hamiltonian systems. J. Diff. Eqs. 11(1972) 109-137. Z 213, 365.

291. Lillo, J. C. On a non-linear differential equation of pendulum type. Thesis. U. of Nebr. 1955.

292. ______, Linear differential equations with almost periodic coefficients. Am. J. Math. 81(1959) 37-45. MR 21, 161, 7.7.

293. ______, On almost periodic solutions of differential equations. Ann. of Math. 69(1959) 467-485. MR 21, 177.

294. ______, Continuous matricies and the stability theory of differential systems. Math. Zeitschr. 73(1960) 45-58. MR 22, 8171, 7.7.

295. ______, A note on the continuity of characteristic exponents. Proc. Nat. Acad. Sci. U. S. A. 46(1960) 247-250. MR 22, 8170.

296. ______, Perturbations of nonlinear systems. Acta. Math. 103(1960) 123-138. MR 22, 1722.

297. ______, Approximate similarity and almost periodic matrices. Proc. Amer. Math. Soc. 12(1961) 400-407. MR 23, A 2433, 7.7.

298. ______, Continuous matrices and approximate similarity. Trans. Amer. Math. Soc. 104(1962) 412-419. MR 26, 5010.

299. Lisevic, L., and Kastjuk, J. Über die Fastperiodizität der Lösungen gewisser gewöhnlichen Differentialgleichungen zweiter Ordnung mit S^p fastperiodischer rechter Seite. Dopovidi Akad. Nauk Ukrain RSR Ser A(1971) 25-26. Z 213, 366.

300. Lisevic, L. Almost-periodic solutions of a hyperbolic system of linear differential equations with almost-periodic coefficients. (Ukrainian). Dopovidi Akad. Nauk Ukr. RSR (1956) 220-222. MR 18, 296.

301. _______, Extension of Favard's theorems to the case of a linear system of differential equations with analytic almost-periodic coefficients. (Ukrainian). Dopovidi Akad. Nauk Ukrain, RSR (1960) 148-149. MR 25, 2289.

302. Ljubarskii, M. The extension of Favard's theory to the case of a system of linear differential equations whose coefficients are unbounded and almost periodic in the sense of Levitan. Soviet Math. 13(1972) 1316-1319.

303. Loonstra, F. Sur les mouvements presque-périodiques. Indagationes Math. 8(1946) 447-454. MR 8, 263.

304. Lyascenko, N. Y. An analogue of the theorem of Floquet for a special case of linear homogeneous systems of differential equations with quasi-periodic coefficients. Dokl. Akad. Nauk SSSR (N. S.) 111(1956) 295-298. (Russian). MR 19, 143.

305. Lykova, O. B. Quasi-periodic solutions of a nearly canonical system. (Russian). Ukrain. Mat. Z. 16(1964) 752-768. MR 32, 7859.

306. Lykova, O. B., and Mitropolsyk, Y. A. Sur l'existence de solutions quasi-périodiques d'un systeme canonique troublé. Colloques Int. CNRS. Les vibrations forcees dans les systemes non-lineaires. (1965) 415-422.

307. Magnus, W. On the exponential solutions of differential equations for a linear operator. Comm. Pure Appl. Math. 7(1954) 649-673.

308. Malkin, I. G. On almost-periodic oscillations of non-linear non-autonomous systems. (Russian). Prikl. Mat. Mech. 18(1954) 681-704, 459-463. MR 16, 590, 249.

309. _______, Some problems of the theory of non-linear oscillations. (Russian). Moscow, 1956. MR 18, 396.

310. Markus, L. Continuous matricies and the stability of differential systems. Math. Zeitschr. 62(1955) 310-319. MR 17, 37, 7.7.

311. Markus, L., and Moore, R. Oscillations and disconjugacy for linear differential equations with almost periodic coefficients. Acta. Math. 96(1956) 99-123. MR 18, 306.

312. Markoff, A. Sur les mouvements presque-périodiques. C. R. 189(1929) 732-734. Fdm. 55, 158.

313. ______, Stabilität im Liapounoffschen Sinne und Fastperiodizität. Math. Z. 36(1933) 708-738. Z 7, 503, 11.11.

314. Massera, J. The existence of periodic solutions of systems of differential equations. Duke Math. J. 17(1950) 457-475. MR 17, 705.

315. ______, Un criterio de existencia de soluciones casiperiodicas de ciertos sistemas de ecuaciones diferenciales casiperiódicas. Publ. Inst. Mat. y Estad. Fac. Ing. y Agrimensura Montevideo 3(1958) 89-102. MR 22, 1709, 6.10.

316. Massera, J. L., and Schäffer, J. J. Linear Differential Equations and functional analysis. I. Ann. Math. 67(1958) 517-572. MR 20, 3466.

317. ______, Linear differential equations and function spaces. Academic Press, New York, 1966.

318. May, L. E. Perturbations in Fully nonlinear systems. SIAM J. Math. Anal. 1(1970) 376-391. MR 42, 2121.

319. Miller, R. K. On almost-periodic differential equations. Bull. Am. Math. Soc. 70(1964) 792-795. MR 29, 4949.

320. ______, Almost periodic differential equations as dynamical systems with applications to the existence of almost-periodic solutions. J. Differential Equations 1(1965) 337-345. MR 32, 2690, 11.11.

321. ______, On asymptotic stability of almost periodic systems. J. Differential Equations 1(1965) 234-239. MR 32, 262.

322. ______, Asymptotic behavior of solutions of nonlinear differential equations. Trans. Amer. Math. Soc. 115(1965) 400-416. MR 33, 7646.

323. ______, Asymptotically almost periodic solutions of a nonlinear Volterra system. SIAM J. Math. Anal. 2(1971) 435-444. MR 45, 5677.

324. ______, Almost-periodic behavior of solutions of a nonlinear Volterra system. Quart. Appl. Math. 28(1971) 553-570. MR 42, 6553.

325. Millionscikov, V. M. Recurrent and almost-periodic limit trajectories of nonautonomous systems of differential equations. (Russian). Doklady Akad. Nauk SSSR 161(1965) 43-44. MR 30, 4043.

326. _______, The structure of fundamental matrices of R-systems with almost periodic coefficients. (Russian). Dokl. Akad. Nauk SSSR 17(1966) 288-291. MR 34, 6196.

327. _______, The connection between the stability of characteristic exponents and almost reducibility of systems with almost periodic coefficients. (Russian). Differencial'nye Uravnenija 3(1967) 2127-2134. MR 36, 6706.

328. _______, Proof of the existence of irregular systems of linear differential equations with almost periodic coefficients. (Russian). Differencial'nye Uravnenija 4(1968) 391-396. MR 37, 5478.

329. _______, A metric theory of linear systems of differential equations. (Russian). Dokl. Akad. Nauk SSSR 179(1968) 20-23. MR 38, 2407.

330. _______, A criterion for the stability of the probable spectrum of linear systems of differential equations with recurrent coefficients and a criterion for the almost reducibility of systems with almost periodic coefficients. (Russian). Dokl. Akad. Nauk SSSR 179(1968) 538-541. MR 38, 2408.

331. _______, A proof of the existence of nonregular systems of linear differential equations with quasiperiodic coefficients. (Russian). Differencial'nye Uravnenija 5(1969) 1979-1983. MR 41, 2127.

332. _______, On the relation between stability of characteristic exponents and almost reducibility of systems with almost periodic coefficients. Differential Equations. 3(1967) 1106-1109 (1972). Z 235, 34108.

333. Minorsky, N. Non-linear oscillations. Van Nostrand, Princeton, N. J., 1962. MR 33, 6061.

334. Misnaevskii, P. On the attainment of almost periodic conditions and almost periodicity of solutions of differential equations in Banach spaces. Vestnik Moskov Univ. Ser I 26(1971) 69-76. Z 227, 34050.

335. Mitropolskii, Y. A., and Samoilenko, A. M. On constructing solutions of linear differential equations with quasiperiodic coefficients by the method of improved convergence. (Russian). Ukrain. Mat. Z. 17(1965) no. 6, 42-59. MR 33, 5992.

336. _______, Quasi-periodic oscillations in Linear systems. Ukr. Math. J. 24(1972) 144-156.

337. Montandon, B. Almost periodic solutions and integral manifolds for weakly nonlinear nonconservative systems. J. diff. Equations 12(1972) 417-425. Z 239, 34019.

338. Moser, J. Combination tones for Duffing's equation. Comm. Pure Appl. Math. 18(1965) 167-181. MR 31, 3678.

339. ______, On the theory of quasiperiodic motions. SIAM Review 8(1966) 145-172. MR 34, 3013.

340. ______, Convergent series expansions for quasi-periodic motions. Math. Ann. 169(1967) 136-176. MR 34, 7888.

341. ______, Perturbation theory for almost periodic solutions for undamped nonlinear differential equations. Int. Symp. Nonlinear Diff. Eq. and Nonlinear Mech. 71-79. Academic Press (1963). MR 26, 5254.

342. ______, On the expansion of quasi-periodic motions in convergent power series. (Russian). Uspehi Mat. Nauk 24(1969) no. 2(146) 165-211. MR 39, 1735.

343. ______, On the construction of almost periodic solutions for ordinary differential equations. Proc. internat. Conf. functional Analysis and related topics. Tokyo 1969, 60-67 (1970). MR 42, 601.

344. Muckenhoupt. C. F. Almost-periodic fuctions and vibrating systems. J. Math. Phys. 8(1928-1929) 163-199. Fdm 55, 759.

345. ______, Vibrating systems and almost periodic functions. Bulletin A. M. S. 35(1929) 178. Fdm 55, 464.

346. Nakajima, F. Existence of almost periodic solutions by Liapunov functions. Japan-U. S. Sem. ordinary differential functional Equations. Kyoto 1971. Lecture Notes Math. 243, 293-297 (1971). Z 226, 34042.

347. ______, Existence of Quasi-periodic solutions of quasi-periodic systems. Funkcial. Ekvac. 15(1972) 61-74.

348. Nasyrov, R. M. On stability of almost-periodic motions in certain critical cases. (Russian). Trudy Univ. Druzby Narodov im. P. Lumumby 5(1964) 30-44. MR 34, 1622.

349. Nemytskii, V. V., and Stepanov, V. V. Qualitative theory of differential equations. Princeton Univ. Press, 1960, VIII + 523 p.

350. Nishida, T. A note on an existence of conditionally periodic oscillation in a one-dimensional anharmonic lattice. Mem. Fac. Engineering Kyoto Univ. 33(1971) 27-34. MR 43, 35109.

351. Obi, C. Uniformly almost periodic solutions of non-linear differential equations of the second order. Proceedings of the International Congress of Mathematicians 1954 volume II. MR 17, 265.

352. ______, Uniformly almost periodic solutions of non-linear differential equations of the second order. I. General exposition. Proc. Cambridge Philos. Soc. 51(1955) 604-613. MR 17, 265.

353. O'Brien, G. C. Almost periodic and quasi-periodic solutions of differential equations. Bull. Austral. math. Soc. 7(1972) 453-454. Z 239, 34018.

354. Oleinik, S. G. The investigation of linear systems of differential equations with almost periodic coefficients. (Russian). Mathematical Physics No. 6 (1969)(Russian), pp. 139-149, Naukova Dumka, Kiev, 1969. MR 41, 8765.

355. Opial, Z. Sur les solutions presque-périodiques des equations differentielles du premier et du second ordre. Ann. Polonici Math. 7(1959) 51-61. MR 22, 799.

356. ______, Sur la stabilite des solutions périodiques et presque-périodiques de l'equation differentielle $x'' + F(x') + g(x) = p(t)$. Bull. Acad. Polon. Sci., Ser. Sci. Math. Astr. Phys. 7(1959) 495-500. MR 22, 2762.

357. ______, Sur les solutions périodiques et presque-périodiques de l'equation differentielle $x'' + kf(x)x' + g(x) = kp(t)$. Ann. Polon. Math. 7(1960) 309-319. MR 22, 5780, 13.6.

358. ______, La presque-periodicité et les trajectoires sur le tore. C. R. Acad. Sci. Paris 250(1960) 3565-3566. MR 22, 8166a.

359. ______, Sur les solutions presque-périodiques d'une classes d'équations différentielles. Ann. Polon. Math. 9(1960/61) 157-181. MR 23, A390, 9.7, 11.11.

360. ______, Sur un theoreme de C. E. Langenhop et G. Seifert. Ann. Polon. Math 9(1960/61) 145-155. MR 22, 12277.

361. ______, Sur une équation différentielle presque-périodique sans solution presque-périodique. Bull. Acad. Polon. Sci. Ser. Sci. Math. Astr. Phys. 9(1961) 673-676. MR 506.

362. Palmieri, G. Soluzioni limitate, o quasi-periodiche, di un'equazione non lineare del calore con discontinuita rispetto all'incognita. (English summary). Ist. Lombardo Accad. Sci. Lett. Rend. A 104(1970) 746-757. MR 45, 5528.

363. Perko, L. M. Higher order averaging and related methods for perturbed periodic and quasi-periodic systems. SIAM J. Appl. Math. 17(1969) 698-724. MR 41, 2129.

364. Perov, A. I. Periodic, almost-periodic, and bounded solutions of the differential equation dx/dt = f(t,x). (Russian). Doklady Akad. Nauk SSSR 132(1960) 531-534. MR 23, A 1084.

365. Perov, A. I., and Kacaran, T. K. Theorems of Favard and Bohr-Neugebauer for multi-dimensional differential equations. (Russian). Izv. Vyss. Ucebn. Zaved. Matematika 1968, no. 5(72) 62-70. MR 37, 4333.

366. Perov, A. I., and Hai, T. K. The almost periodic solutions of homogeneous differential equations. (Russian). Differencial'nye Uravnenija 8(1972) 453-458. MR 45, 5489.

367. Petersen, R. On the Laplace transformation of an almost periodic function. Festskrift til Professor, Dr. Phil. J. F. Steffensen fra Kolleger og Elever paa hans 70 Aars Fodselsdag 28. Februar 1943, pp. 113-139. Den Danske Aktuarforening, Copenhagen, 1943. (Danish). MR 8, 152.

368. ______, Laplace-transformation of almost-periodic functions. Den 11-te Skandinaviske Matematikerkongress, Trondheim, 1949, 158-165. MR 14, 746.

369. Prouse, G. Soluzioni quasi-periodiche delle equazioni lineari di tipo parabolico. Rend. 1st. Lomb. (A) 96(1962) 847-860. MR 27, 6042.

370. ______, Soluzioni quasi-periodiche di un'equazione iperbolica a coefficienti cariabili. Ric. de Mat. XI(1962) 123-138. MR 35, 1920.

371. ______, Sulle equazioni differenziali astratte quasi-periodiche secondo Stepanoff. Ric. di Mat. XI(1962) 254-270. MR 35, 1921.

372. ______, Soluzioni quasi-periodiche dell'equazione differenziale di Navier-Stokes in due dimensioni. Rend. Sem. Mat. Padova 33(1963) 186-212. MR 30, 779.

373. ______, Soluzioni quasi-periodiche dell'equazione non omogenea della membrana vibrante, con termine dissipativo quadratico. Rend. Acad. Naz. Lincei 37(1964) 246-252. MR 33, 4492.

374. ______, Soluzioni quasi-periodiche dell'equazione non omogenes delle onde con termine dissipativo non linear. I, II, III, IV. Rend. Accad. Naz. Lincei 38(1965) 804-807; 39(1965) 11-18, 155-160, 240-244. MR 35, 1922.

375. ______, Periodic or almost-periodic solutions of a non-linear functional equation, I, II, III, IV. Rend. Acc. Naz. Lincei (1967-68) 43-44. MR 40, 6093-6095.

376. Putnam, C. R. Stability and almost-periodicity in dynamical systems. Proc. Am. Math. Soc. 5(1954) 352-356. MR 18, 958.

377. ______, Unilateral stability and almost periodicity. J. Math. Mech. 9(1960) 915-917. MR 22, 9668.

378. Ragimov, M. B., and Zadoroznii, V. G. Almost periodic solutions of multidimensional differential equations. (Russian Azerbaijani summary). Akad. Nauk Azerbaidzan. SSR Dokl. 26(1970) no. 9, 8-11. MR 45, 662.

379. Rao, A. On differential Operators with Bohr-Neugebauer Type Property. J. Differential equations 13(1973) 490-494.

380. Razin, G. A. The almost-periodicity of the solutions of a certain class of systems of stochastic differential equations. (Russian). Differencial'nye Uravnenija 7(1971) 1419-1428. 1540. MR 45, 5505.

381. Ricci M. L. Sulle equazioni differenziali con termine noto quasi-periodico secondo Stepanoff. Rend. Ist. Lombardo Sci. Lettere 96(1962) 861-882. MR 27, 5970.

382. ______, Soluzioni quasi-periodiche di un'equazione a derivate parziali del tipo di Fuchs. Ann. Mat. Pura Appl. 79(1966) 367-404. MR 35, 1923.

383. ______, Sull'integrazione con la trasformata di Fourier di un'equazione a derivate parziali del tipo di Fuchs. Ist. Lombatdo Accad. Sci. Lett. Rend. A 101(1967) 29-48. MR 37, 3219.

384. Ricci, M. L., and Rizzonelli, P. Alcuni complementi alla teoria delle funzioni quasi-periodiche astratte. Ist. Lombardo Accad. Sci. Lett. Rend. A 95(1961) 525-534. MR 25, 2381.

385. ______, Sulle funzioni 1^1-quasi-periodiche. Ist. Lombardo Accad. Sci. Lett. Rend. 95(1961) 941-946. MR 26, 4120.

386. Rjabov, J. A. On a method of finding a bound for the region of existence of periodic and almost-periodic solutions of quasi-linear differential equations with a small parameter and a non-linear characteristic of non-linearity. (Russian). Izv. Vyss. Ucebn. Zaved. Matematika nr. 2(33) (1963) 101-107. MR 27, 3876.

387. Roettinger, I. Note on use of almost-periodic functions in the solution of certain boundary value problems. J. Math. Phys. 27(1948) 232-239. MR 10, 197.

388. Romanov, V. I. Application of the Poincare-Andronov operator to quasiperiodic solutions of a certain class of nonlinear systems of ordinary differential equations. (Russian). Equations of Math. Phys. and Funct. Anal. (Russian). (1966) 144-148. MR 35, 470.

389. ______, The reducibility of certain systems with periodic and quasiperiodic matrices. (Russian). Izv. Vyss. Ucebn. Zaved. Matematika 1969, no. 7 (86) 70-73. MR 40, 3033.

390. Romanov, V. I., and Harashal, V. H. Reducibility of linear systems of differential equations with quasi-periodic coefficients. (Russian). Differencial'nye Uravnenija 2(1966) 1423-1427. MR 34, 6197.

391. Sabirov, T. Estimates of the Green's function of a differential operator with a small parameter and almost periodic coefficients. (Russian. Tajiki summary). Dokl. Akad. Nauk Tadzik. SSR 14(1971) no. 6, 14-17. MR 45, 2230.

392. Samiolenko, A. M. Reducibility of a system of ordinary differential equations in a neighborhood of a smooth integral manifold. (Russian). Ukrain. Mat. Z. 18(1966) no. 6, 41-64. MR 35, 4494.

393. ______, The reducibility of systems of linear differential equations with quasiperiodic coefficients. (Russian). Ukrain. Mat. Z. 20(1968) 279-281. MR 37, 504.

394. Sansone, G. Equazioni differenziali nel campo reale. I, II. (Seconda edizione). Bologna, 1948-1949.

395. Sandor, S. T. Asupra ecuatiilor diferentiale liniare de ordin superior cu coeficietni aproape-periodici. Bul stiint. Acad. R. P. R. 7(1955) 329-346. MR 17, 613.

396. ______, Ecuatiile diferentiale liniare neomogene cu coeficienti aproape-periodici si ecuatiile cvasiliniare cu parametru mic. Bul. Stiint Acad. R. P. R. 7(1955) 623-698. MR 17, 614.

397. Schaeffer, A. J. Boundedness of solutions to linear differential equation. Bull. Amer. Math. Soc. 74(1968) 508-511. MR 36, 5441.

398. Schaffer, J. J. Analytische Parameterabhängigkeit der fastperiodischen Lösungen. Rend. Circ. Mat. Palermo 5(1956) 204-236. MR 18, 576.

399. ______, Sobre ecuaciones diferenciales nolineales casi-periodicas. Publ. Inst. Mat. Estad. Fac. Ing. Agr., Montevideo 3(1957) 17-52. MR 20, 1033.

400. ______, Linear differential equations and functional analysis. IX. Almost-periodic equations. Math. Ann. 150(1963) 111-118. MR 27, 5961.

401. Scharf, G. Fastperiodische Potentiale. Buchdrucherei Birkhauser AG Basel (1965) 34 pp. MR 34, 1603.

402. Seifert, G. A rotated vector approach to the problem of stability of solutions of pendulum-type equations. Cont. to the Theory of non-Linear Oscillations - S. Lefschetz, ed. Princeton U. Press (1956) 1-17. MR 26, 6500.

403. ______, Almost-periodic solutions for systems of differential equations near points of nonlinear first approximations. Proc. Am. Math. Soc. 11(1960) 429-435. Mr 26, 6500.

404. ______, Stability conditions for separation and almost-periodicity of solutions of differential equations. Contr. Differential Equations 1(1963) 483-487. MR 26, 6501.

405. ______, Uniform stability of almost-periodic solutions of almost-periodic systems of differential equations. Contr. Differential Equation II(1963) 269-276. MR 27, 5987.

406. ______, A condition for almost periodicity with some applications to functional-differential equations. J. Differential Equation 1(1965) 393-408. MR 31, 4962, 9.7.

407. ______, Stability conditions for the existence of almost-periodic solutions of almost-periodic systems. J. Math. Anal. Appl. 10(1965) 409-418. MR 30, 3276.

408. _______, Some recent results for almost periodic solutions of differential equations. SIAM Review 7(1965) 524-538. MR 32, 4331.

409. _______, Almost-periodic solutions for almost-periodic systems of ordinary differential equations. J. Differential Equations 2(1966) 305-319. MR 34, 4613.

410. _______, Stability and uniform stability in almost periodic systems. Proceedings United States-Japan Seminar on Differential and Functional Equations. (1967) 577-580.

411. _______, On total stability and asymptotic stability. Tohoku Math. J. (2) 19(1967) 71-74. MR 35, 5732.

412. _______, Almost periodic solutions and Asymptotic stability. J. Math. Anal. Appl. 21(1968) 136-149. MR 36, 1774, 11.11.

413. _______, Recurrence and almost periodicity in ordinary differential equations. Advances in Differential and Integral Equations. SIAM (1969) 119-121.

414. _______, Almost periodic solutions by the method of averaging. Lecture Notes 243(1971), Japan-United States Seminar on Ordinary Differential and Functioanl Equations, 123-133.

415. _______, On almost periodic solutions for undamped systems with almost periodic forcing. Proc. Amer. Math. Soc. 31(1972) 104-108. MR 44, 2989.

416. _______, Almost periodic solutions for limit periodic systems. SIAM J. Appl. Math. 22(1972) 38-44.

417. Sell, G. Nonautonomous Differential Equations and Topological Dynamics. I. The Basic Theory. II. Limiting Equations. Trans. Amer. Math. Soc. 127(1967) 241-262, 263-283. MR 35, 3187, 11.11.

418. _______, Some perturbation problems in ordinary differential equations. Funkcial Ekvac. 10(1967) 1-13. MR 36, 5456, 11.11.

419. _______, Invariant measures and Poisson stability, Topological Dymanics, Benjamin, 1968, 435-454.

420. _______, Topological dynamics and ordinary differential equations. (Van Nostrand Reinhold Mathematical Studies, 33) London etc.: Van Nostrand Reinhold Company 1971, IX, 199 p. £ 2.10. Z 212, 292.

421. _______, Almost Periodic Solutions of Linear Partial Differential Equations. J. Math. Anal. Appl. 42(1973) 302-312.

422. _______, A note on almost periodic solutions of linear partial differential equations. Bull. Amer. Math. Soc. 79(1973) 428-430.

423. Sethna, P. R. Domains of attraction of some quasiperiodic solutions. Differential Equations and Dynamical Systems (1967) 323-332. Academic Press. MR 36, 2897.

424. _______, An extension of the method of averaging. Quart. Appl. Math. 25(1967) 205-211. MR 36, 474.

425. Sethna, P. R., and Moran, T. J. Some nonlocal results for weakly nonlinear dynamical systems. Quart. Appl. Math. 26(1968) 175-185. MR 39, 1763.

426. Sibuya, Y. Sur les solutions borneés d'un système des équations différentielles ordinaires non-linéaires e coefficients périodiques. J. Fac. Sci. Univ. Tokyo 7(1956) 333-341. MR 27, 5969.

427. _______, Sur un système d'équations différentielles ordinaires à coefficients presque-périodiques et contenant des paramètres. J. Fac. Sci. Univ. Tokyo 7(1957) 407-417. MR 19, 141.

428. _______, On bounded solutions of ordinary differential equations with almost-periodic coefficients. Bol. Soc. Mat. Mexicana (2) 5(1960) 290-293. MR 25, 3227.

429. _______, Almost-periodic solutions of a system of ordinary differential equations with periodic coefficients. Natur. Sci. Rep. Ochanomizu Univ. 13(1962) no. 2, 21-30. MR 25, 3227.

430. _______, Almost periodic solutions of Poisson's equation. Proc. Amer. Math. Soc. 28(1971) 195-198. MR 44, 3033.

431. Simanov, S. N. Almost-periodic solutions of nonhomogeneous linear differential equations with lag. (Russian). Izv. Vyss. Ucebn. Zaved. Matematika (1958) no. 4(5) 270-274. MR 24, A 1470.

432. _______, Almost-periodic solutions in non-linear systems with retardation. (Russian). Doklady Akad. Nauk SSSR 125(1959) 1203-1206. MR 21, 5051.

433. _______, The degenerate case of almost periodic oscillations of quasi-linear systems with time delay. (Russian). Dokl. Akad. Nauk SSSR 133(1960) 36-39. (Translated - Soviet Math. Dokl. 1(1961) 812-815. MR 24, A 1492.

434. Sircenko, Z. F. On the existence and properties of an almost periodic solutions in standard form in a Hilbert space near an equilibrium point. (Ukranian. Russian and English summary). Dopovidi Akad. Nauk Ukrain. RSR (1964) 1132-1135. MR 29, 6115.

435. Smolickii, K. L. On almost-periodicity of generalized solutions of wave equation. (Russian). Doklady Akad. Nauk SSSR 60(1948) 353-356. MR 9, 513.

436. Smulev, I. I. Almost periodic and periodic solutions of the problem with oblique derivative for parabolic equations. (Russian). Differencial'nye Uravnenija 5(1969) 2225-2236. MR 42, 648.

437. _______,Bounded, periodic and almost periodic solutions of elliptic equations. (Russian). Differencial'nyc Uravnenija 7(1971) 519-529. MR 44, 4342.

438. Sobolev, S. L. Sur la presque-périodicité des solutions de l'équation des ondes, I, II, III. Comptes Rendus (Doklady) de l'Acad. Sci. de l'U. R. S. S. 48(1945) 542-545; 618-620; 49(1945) 12-15. MR 8, 78.

439. Sprindzhuk, V. Asymptotic behavior of the integrals of quasiperiodic functions. Diff. Uraveniya 3(1967) 862-868. MR 35, 3339.

440. Stelik, V. G. On the solutions of a linear systems of differential equations with almost periodic coefficients. Ukrain Math. Z. 10(1958) 318-327. MR 21, 2087.

441. Stepanoff, W. Sur le probleme de M. Levi-Civita concernant le mouvement moyen. Atti. Accad, naz. Lincei Rend. VI. s. 17(1933) 526-531. Zbl. 7, 179.

442. Sternberg, S. Celestial Mechanics parts I, II. (1969).

443. Stokalo, I. Z. Criteria for stability and instability of the solutions of linear differential equations with quasi-periodic coefficients. Akad. Nauk URSR· Informaciinii Byuleten no. 1(10-11) 38-39 (1945). (Ukrainian and Russian) (Levinson) MR 7, 520.

444. _______, Linear differential equations of the nth order with quasiperiodic coefficients. Akad. Nauk URSR. Informaciinii Byuleten no. 1(10-11) 40-42 (1945). (Ukrainian and Russian) (Levinson). MR 7, 520.

445. _______, Systems of linear differential equations with quasiperiodic coefficients. Akad. Nauk URSR. Informaciinii Byuleten. no. 1(10-11) 42-45 (1945). (Ukrainian and Russian)(Levinson). MR 7, 520.

446. _______, A stability and instability criteria for solutions of linear differential equations with quasi-periodical coefficients. Rec. Math. N. S. 19(61) 263-286 (1946). (Russian. English summary)(Bellman). MR 8, 329.

447. _______, Linear differential equations of the nth order with quasi-periodical coefficients. Rep. Acad. Sci. Ukrainian SSR no. 3-4, 17-20 (1946). (Ukrainian. Russian and English summaries)(Levinson) MR 8, 464.

448. _______, Systems of linear differential equations with quasi-periodical coefficients. Rep. Acad. Sci. Ukrainian SSR no. 3-4, 21-24 (1946). (Ukrainian. Russian and English summaries)(Levinson). MR 8, 464.

449. _______, Stability and instability criteria for the solutions of linear differential equations with quasi-periodic coefficients. (Russian). Mat. Sb. 19(1946) 263-286. MR 8, 329.

450. _______, On the theory of linear differential equations with quasi-periodic coefficients. (Russian). Sb. Trudov Inst. Mat. Akad. Nauk USSR 8(1946).

451. _______, On the theory of linear differential equations with quasiperiodic coefficients. Akad. Nauk Ukrain. RSR. Zbirnik Prac. Inst. Mat. 1946, no. 8, 163-176 (1947). (Ukrainian. Russian and English summaries)(Levinson). MR 12, 334.

452. _______, Theory of symbolic representation of solutions of linear differential equations with quasi-periodic coefficients. (Russian). Sb. Trudov Inst. Mat. Akad. Nauk USSR 11(1948) 43-59.

453. _______, Linear Differential Equations with Variable Coefficients. Gordon-Breach Publ. Co. translated from the Russian, 1961. MR 27, 5979.

454. Svinbekov, K. D. On the analytic form of solutions of linear systems of differential equations with quasi-periodic coefficients. (Russian). Izv. Akad. Nauk Kazah. SSR Ser. Fiz. Mat. Nauk (1964), no. 2, 69-71. MR 30, 3263.

455. Taam, C. T. Stability periodicity, and almost periodicity of solutions of nonlinear differential equations in Banach spaces. J. Math. Mech. 15(1966) 849-876. MR 33, 4433.

456. Takahashi, S. On the formal integration of the Fourier series of an almost-periodic function. Mat. Tidsskrift (1935) 82-85. Z 13, 112.

457. ______, An application of the Fourier transform to almost-periodic functions. Proc. Imp. Acad. Tokyo 14(1939) 87-89. Z 19, 205.

458. Talpalaru, P. Solutions periodiques et presque-périodiques des systems differentiels. An. sti. Univ. Al. I. Cunza Iasi n Ser Sect. Ia. 15(1969) 375-385. MR 42, 2107.

459. Tazimuratov, I. On the existence of almost periodic solutions to systems close to those of Liapunov. (Russian). Izv. Akad. Nauk Kazah. SSR Ser. Fiz. Mat. Nauk (1966) no. 1, 102-106. MR 34, 3014.

460. ______, Existence of quasi-periodic solutions of Liapunov systems. (Russian). Equations of Math. Phys. and Funct. Anal. (Russian), pp. 154-159. Izdat. "Nauka" Kazah. SSR, Alma-Ata, 1966. MR 34 6252.

461. Terras, R. On almost periodic and almost automorphic differences of functions. Duke Math. J. 40(1973) 81-92.

462. Tornehave, H. A theorem on the mean motions of almost-periodic functions. Mat.-fysiske Medd. 25 20(1950) 1-18. MR 12, 22.

463. ______, Recent investigations on almost-periodic movements. Tolfte Scandinaviska Matematiker-kongressen, Lund. 1953, 302-309. MR 16, 610.

464. ______, On almost-periodic movements. Mat.-fysiske Medd. 28 13(1954) 1-42. MR 16, 735.

465. Tulegenov, B. On the existence of quasi-periodic solutions of a non-linear first-order differential equation. (Russian). Izv. Akad. Nauk Kazah. SSSR. Ser. Fiz. mat. nauk no. 2(1964) 72-76. MR 30, 3264.

466. Turcu, A. Solutii aproape-periodice ale ecuatiei lui Duffing in caz de nerezonanta. Studia Univ. Babes-Bolyai, Ser. Math.-Phys. 9, No. 2(1964) 61-76. MR 31, 433.

467. _______, Almost-periodic solutions of the equation of Duffing in the case of resonance. Studia Univ. Babes-Bolyaik Ser. Math.-Phys. 10(1965) no. 1, 83-94. MR 32, 2685.

468. Umbetzanov, D. U. On the question of periodic and almost periodic solutions of certain quasi-linear differential equations. (Russian). Izv. Akad. Nauk Kazah. SSR Ser. Fiz.-Mat. Nauk (1965) no. 1, 54-64. MR 32, 2678.

469. _______, An existence theorem for almost periodic solution of differential equations in linear normed spaces. (Russian). Izv. Akad. Nauk Kazah. SSR Ser. Fiz.-Mat. Nauk (1967) no. 1, 38-44. MR 35, 1904.

470. _______, Periodic and almost periodic solutions, of a certain class of partial differential equations, which are holomorphic with respect to a parameter. (Russian. English summary). Vestnik Moskov. Univ. Ser. I Mat. Meh. 24(1969) no. 3, 37-43. MR 41, 2193.

471. _______, The almost periodic solutions of certain class of partial differential equations with small parameters. (Russian). Differencial'nye Uravnenija 6(1970) 913-916. MR 42, 6405.

472. _______, The almost-periodic solutions of a certain system of equations of hyperbolic type. (Russian. Kazah summary). Izv. Akad. Mauk Kazah. SSR Ser. Fiz.-Mat. 1971, no. 3, 55-59. MR 45, 5533.

473. Urabe, M. Green functions of pseudoperiodic differential operators. Lecture Notes 243(1971) Japan-United States Seminar on Ordinary Differential and Functional Equations 106-122.

474. _______, Existence Theorems of Quasiperiodic solutions to nonlinear differential systems. Funk. Ekv. 15(1972) 75-100.

475. Vaghi, C. Sulla regolarizzazione delle soluzioni quasi-periodiche dell'equazione non omogenea delle onde. Ist. Lombardo Accad. Sci. Lett. Rend. A 96(1962) 267-285. MR 26, 6619.

476. _______, Soluzioni C-quasi-periodiche dell'equazione non omogenea delle onde. Ric. Mat. 12(1963) 195-215. MR 29, 3753.

477. _______, Su un'equazione iperbolica con coefficienti periodici e termine noto quasi-periodico. Rend. Ist. Lomb. (A) 100(1966) 155-180. MR 35, 568.

478. ______, Soluzioni limitate, o quasi-periodiche, di un'equazione di tipo parabolico non lineare. Boll. U. M. I. 4-5(1968) 559-580. MR 39, 3134.

479. ______, Su un'equazioni parabolica con termine quadratico nella derivata z_x. Rend. Ist. Lombardo 104(1970) 3-23. Z 202, 113.

480. Valeev, K. Linear differential equations with quasi-periodic coefficients and constant retardation of the argument. Visnik Kiev Univ. (1969) Ser. Mat. Mech. 16-24. MR 43, 2329.

481. Van Vleck, F. S. A Note on the relation between periodic and orthogonal fundamental solutions of linear systems. Amer. Math. Monthly (71) 1964, 406-408. MR 28, 5217.

482. ______., A Note on the relation between periodic and orthogonal fundamental solutions of linear systems. Amer. Math. Monthly (71) 1964, 774-776. MR 31, 6029-6030.

483. Vasconi, A. Sull'integrazione delle funzioni quasi-periodiche secondo Stepanoff negli spazi di Clarkson. Rend. Ist. Lombardo Sci. Lettere 95(1961) 1024-1029. MR 35, 1799.

484. ______, Sull'equazione delle onde con termine noto quasi-periodico secondico Stepanoff. Ist. Lombardo Accad. Sci. Lett. Rend. A 96(1962) 903-914. MR 28, 338.

485. Veech, W. A. Almost automorphic functions on groups. Amer. J. Math 87(1965) 719-751. MR 32, 4469.

486. Vereennilov, V. On the construction of solutions of quasilinear nonautonomous systems in resonance cases. J. Appl. Math. Mech. 33(1969) 1090-1099. MR 42, 2108.

487. Vieth, W. Bounded Banach-valued functions with almost periodic differences. Boll. U. M. I. 4(1971) 153-163.

488. Vzovskii, D. A. The almost periodic solutions of certain nonlinear systems of differential equations with deviating argument. (Russian). Differencial'nye Uravnenija 8(1972) 415-423. MR 45, 7224.

489. Walther, A. Über lineare Differenzengleichungen mit konstanten Koeffizienten und fastperiodischer rechter Seite. Nachr. Ges. Wiss. Gottingen. Math.-Phys. Klasse (1927) 196-216. Fdm 53, 435.

490. Wei, J., and Norman, E. On global representations of the solutions of linear differential equations as a product of exponentials. Am. Math. Soc. Proc. 15(1964) 327. MR 28, 3223.

491. Weinstein, A. Sulle soluzioni quasi-periodiche di una classe di equazioni ellittiche. Rend. Accad. Naz. Lincei (8) 32(1962) 863-866. MR 26, 5274.

492. Wexler, D. Solutions périodiques et presque-périodiques des systemes d'équations différentielles aux impulsions. C. R. Acad. Sci. Paris 259(1964) 287-289. MR 29, 2471.

493. ______, Solutions périodiques et presque-périodiques des systèms d'équations différentielles aux impulsions. Rev. Roumaine Math. Pures Appl. 10(1965) 1163-1169. MR 35, 6932.

494. ______, Solutions presque-périodiques des systèmes d'équations différentielles a perturbations distributions. C. R. Acad. Sci. Paris 262(1966) A 436-439. MR 32, 7867.

495. ______, Solutions périodiques et presque-périodiques des systèmes d'équations différentielles linéaires en distributions. Journal Diff. Equations 2(1966) 12-32. MR 33, 354.

496. ______, Solutions presque-périodiques des systèmes linéaires a perturbation-distribution. Rev. Roumaine Math. Pures Appl. 13(1968) 111-129. MR 40, 463.

497. Weyl, H. Integralgleichungen und fastperiodische Funktionen. Math. Ann. 97(1927) 338-356.

498. ______, Mean Motion. Am. J. Math. 60(1938) 889-896, 61(1939) 143-148.

499. Wiener, N., and Wintner, A. On the ergodic dynamics of almost periodic systems. Am. J. Math. 63(1941) 794-824. MR 4, 15.

500. Willett, D. Uniqueness for second order nonlinear boundary value problems with applications to almost periodic solutions. Ann. Mat. Pura. Appl. 81(1969) 77-92. MR 40, 444.

501. Wintner, A. Über einige Anwendungen der Theorie der fastperiodischen Funktionen auf das Levi-Civitasche Problem der mittleren Bewegung. Annali di Mat. pura ed appl. 10(1932) 277-282. Z 5, 30.

502. _______, On an application of diophantine approximation to the repartition problems of dynamics. J. London Math. Soc. 7(1932) 242-246. Z 5, 405.

503. _______, On the asymptotic differential distribution of almost-periodic and related functions. Am. J. Math. 56(1934) 401-406. Z 9, 253.

504. _______, Almost-periodic functions and Hill's theory of lunar perigee. Am. J. Math. 59(1937) 795-802. Z 18, 95.

505. _______, Liouville systems and almost periodic functions. Amer. J. Math. 60(1938) 463-472. Z. 20, 172.

506. _______, On the almost-periodic behavior of the lunar node. Am. J. Math. 61(1939) 49-60. MR 1, 124.

507. _______, On an ergodic analysis of the remainder term of mean motions. Proc. Nat. Acad. Sci. U. S. 26(1940) 126-129. MR 1, 147.

508. Yartsev, A. Linear Integro-differential equations with almost periodic coefficients. Math. Notes 8(1970) 729-735.

509. Yoshizawa, T. Extreme stability and almost-periodic solutions of functional-differential equations. Arch. Rat. Mech. Anal. 17(1964) 149-170. MR 29, 3734.

510. _______, Asymptotic stability of solutions of an almost-periodic system of functional-differential equations. Rend. Circ. Math. Palermo (2) 13(1964) 209-221. MR 32, 2693.

511. _______, Stability theory by Liapunov's second method. Chap. VII. Math. Soc. Japan, Tokyo, 1966. MR 34, 7896, 11.11

512. _______, Existence of a globally uniform-asymptotically stable periodic and almost periodic solution. Tohoku Math. J. (2) 19(1967) 423-428. MR 37, 525, MR 38, 2387.

513. _______, Stability and existence of periodic and almost periodic solutions. Proceedings United States-Japan Seminar on Differential and Functional Equations. (1967) 411-428. MR 36, 5448.

514. _______, Some remarks on the existence and the stability of almost periodic solutions. Advances in Differential and Integral Equations. SIAM (1969) 166-174.

515. _______, Asymptotically almost periodic solutions of an almost periodic system, Funkcial. Ekvac. 12(1969) 23-40, MR 41, 2132, 11.11.

516. _______, Stability for almost periodic systems. Lecture Notes 243(1971) Japan-United States Seminar on Ordinary Differential and Functional Equations, 29-39. Z 226, 34043.

517. Zaicev, A. I. Almost periodic solutions of linear systems of ordinary differential equations. (Russian). Differential Equations and their Appl. (1967) 35-39. MR 36, 6708.

518. _______, Analytic form of solutions of linear systems of differential equations with quasi-periodic coefficients. (Russian). Differencial'nye Uravnenija 3(1967) 219-225. MR 35, 1873.

519. Zaicev, A., and Kapish, K. Quasiperiodic solutions of nonlinear systems of differential equations containing a small parameter. Diff. Eqs. 7(1967) 1100-1112. MR 35, 6947.

520. Zaidman, S. Sur la presque-périodicite des solutions de l'equation des ondes non homogene. J. Math. Mech. 8(1959) 369-382. MR 21, 2812.

521. _______, Solutions presque-périodiques des equations hyperboliques. C. R. Acad. Sci. Paris 250(1960) 2112-2114. MR 22, 8211.

522. _______, Solutions presque-périodiques dans le probleme de Cauchy, pour l'equation non-homogene des ondes. I, II. Rend. Accad. Naz. Lincei (8) 30(1961) 677-681, 823-827. MR 24, A 3415.

523. _______, Soluzioni limitate e quasi-periodiche dell'equazione del calore non-omogenea. I, II. Rend. Accad. Naz. Lincei (8) 31(1961) 362-368; (8) 32(1962), 30-37. MR 26, 6609.

524. _______, Solutions presque-périodiques des equations hyperboliques. Ann. Sci. Ecole Norm. Sup. Paris (3) 79(1962) 151-198. MR 27, 2723.

525. _______, Teoremi di quasi-periodicita per alcune equazioni differenziali operazionali. Rend. Sem. Mat. Fis. Milano 33(1963) 220-235. MR 28, 1369.

526. _______, Quasi-periodicita per l'equazione di Poisson. Rend. Accad. Naz. Lincei (8) 34(1963) 241-245. MR 28, 5262.

527. _______, Quasi-periodicita per una equazione operazionale del primo ordine. Rend. Accad. Naz. Lincei 35(1963) 152-157. MR 29, 5134.

528. _______, Soluzioni quasi-periodiche per alcune equazioni differenziali in spazi Hilbertiani. Ric. mat. 13(1964) 118-134. MR 29, 3910.

529. _______, Almost-periodicity for some differential equations in Hilbert spaces. Rend. Accad. Naz. Lincei (8) 37(1964) 253-257. MR 32, 2710.

530. _______, Soluzioni limitate o quasi-periodiche dell'equazione di Poisson. Ann. di Mat. pura ed appl. 64(1965) 365-405. MR 32, 6037.

531. _______, Spectrum of Almost periodic solutions for some abstract differential equations. J. Math. Anal. Appl. 28(1969) 336-338. MR 39, 7303.

532. _______, Some asymptotic theorems for abstract differential equations. Proc. AMS 25(1970) 521-515. MR 42, 640.

533. Zikov, V. V. Abstract equations with almost periodic coefficients. (Russian). Dokl. Akad. Nauk SSSR 163(1965) 555-558. MR 32, 6014.

534. _______, Almost periodic solutions of differential equations in Hilbert space. (Russian). Dokl. Akad. Nauk SSSR 165(1965) 1227-1230. MR 33, 355.

535. _______, On the existence of almost periodic solutions of operational differential equations. (Russian). Collection of scientific papers, Polytechnic of Vladmir, 1969.

536. _______, Almost periodic solutions of linear and nonlinear equations in a Banach space. Soviet Math. 11(1970) 1457-1461. MR 43, 667.

537. _______, Die Existenz nach Levitan fastperiodischer Lösungen linearer systems. Mat. Zametki 9(1970) 409-414. Z 224, 34036.

538. _______, The existence of Levitan almost-periodic solutions of linear systems (second complement to Favard's classical theory). Math. Notes 9, 235-238 (1971). Z 235, 34095.

539. _______, Remarks on compactness conditions related to the work of M. I. Kudets on integration of abstract almost-periodic functions. Func. Anal. Appl. 5(1971) 26-30.

540. Zmûro, V. Reduction of a problem on stability of solutions for a differential equation with almost periodic coefficeints to multi-dimensional determinants. Dopovidi Akad. Nauk Ukrain. RSR SerA. (1971) 256-259. Z 219, 193.

541. Zubov. V. I. On almost periodic solutions of systems of differential equations. (Russian). Vestnik Leningrad. Univ. 15(1960) No. 1, 104-106. MR 22, 3848.

542. __________, Periodic and almost-periodic forced oscillations arising from the action of an external force. (Russian). Izv. Vyss. Ucebn. Zaved. Mat. No. 3(19) (1960) 93-102. MR 25, 1335.

543. __________, Oscillations in non-linear and control systems. (Russian). Leningrad, 1962. MR 32, 5954.

Additional References

544. Corduneanu, C. Almost periodic functions, Interscience, New York, 1968.

545. Hahn, W. Stability of Motion, Springer-Verlag, 1967.

546. Jacobs, K. Einige Grundbegriffe der Topologischen Dynamik, math-phys. Semesterberichte XIV, 1967, 129-150.

547. Meisters, G. H. On the almost periodicity of the integral of an almost periodic function, Amer. Math. Soc. Notices, 5(1958), 683.

548. Roseau, M. Vibrations non linéares et théorie de la stabilité. Springer-Verlag, 1966. MR 33, 51-71.

1. Notations:

$\| \ \|$	supremum norm, see 1.2.
C	complex numbers, see 1.2.
R	real numbers, see 1.2.
H(f)	hull of f , see 1.3.
AP(C)	complex valued a.p. functions, see 1.2.
AP(E)	real or complex valued a.p. functions, see 1.2.
$\beta \subset \alpha$	βeta is a subsequence of α , see 1.2.
$T_\alpha f$	$T_\alpha f(x) = \lim_n f(x+\alpha_n)$, see 1.3.
$T(f,\epsilon)$	ϵ-translation numbers of f , see 1.7.
E^n	real or complex n-space, see 2.9.
$a(f,\lambda)$	λ th fourier coefficient of f , see 3.2.
$a(f,\lambda,x)$	λ th fourier coefficient of $f(x,\cdot)$, see 3.2.
exp(f)	fourier exponents of f , see 3.7.
mod(f)	module of f , see 4.5.
$M_n, AP_n, AT_n, K_n, T_n, R_n, AR_n$	matrix classes, see 7.15, 7.16.
$T_\alpha(id) = \infty$	$\lim_n \alpha_n = \infty$, see 9.2.

2. General index